Air Pollution, Acid Rain and the Environment

Watt Committee Report Number 18

The Watt Committee on Energy

AIR POLLUTION, ACID RAIN AND THE ENVIRONMENT
MEMBERS OF WORKING GROUP AND SUB-GROUPS

Dr Helen ApSimon, *Imperial College of Science and Technology, London*

Dr W.V.C. Batstone, *Milton Keynes*

Dr. R.W. Battarbee, *University College, London*

Dr J.N.B. Bell, *Imperial College of Science and Technology, London*

J. Bernie, *Paint Research Association, Teddington*

Dr W.O. Binns, *Forestry Commission, Farnham*

Dr R.N. Butlin, *Building Research Establishment, Garston*

T. Carrick, *Freshwater Biological Association, Ambleside*

Dr M.J. Chadwick, *University of York*

A.J. Clarke, *Central Electricity Generating Board, London*

Dr Alan Cocks, *Central Electricity Research Laboratories, Leatherhead*

Dr M.J. Cooke, *Coal Research Establishment, Stoke Orchard*

Prof. R.U. Cooke, *University College, London*

Dr David Cope, *UK Centre for Economic and Environmental Development, London*

Dr E. Cowell, *BP, London*

D.H. Crawshaw, *North West Water Authority, Warrington*

C.J. Davies, *National Coal Board, Harrow*

Dr G. Dollard, *Atomic Energy Research Establishment, Harwell*

Dr W.M. Edmunds, *British Geological Survey, Wallingford*

Dr B. Fisher, *Central Electricity Research Laboratories, Leatherhead*

M.J. Flux, *Imperial Chemical Industries plc, London*

Dr G.B. Gibbs, *Central Electricity Research Laboratories, Leatherhead*

D. Hammerton, *Clyde River Purification Board, East Kilbride*

Dr R. Harriman, *Department of Agriculture and Fisheries for Scotland, Pitlochry*

Dr J.E. Harris, *Berkeley Nuclear Laboratories, CEGB*

Dr N.H. Highton, *Monopolies and Mergers Commission, London*

A.V. Holden, *Pitlochry*

Dr G.D. Howells, *Central Electricity Generating Board, Leatherhead*

Dr J.B. Johnson, *Corrosion and Protection Centre, University of Manchester, Institute of Science and Technology*

P. Jones, *Institute of Petroleum, London*

A.S. Kallend, *Central Electricity Research Laboratories, Leatherhead*

Dr A.W.C. Keddie, *Department of Trade and Industry, London*

Dr J.A. Lee, *Victoria University of Manchester*

Dr G. Lloyd, *National Physical Laboratory*

Dr P.S. Maitland, *North of Scotland Hydro Electric Board, Sloy*

Dr M.I. Manning, *Central Electricity Research Laboratories, Leatherhead*

Prof. T. Mansfield, *University of Lancaster*

A.R. Marsh, *Central Electricity Research Laboratories, Leatherhead*

C. Martin, *L.G. Mouchel and Partners, Bath*

Prof. K. Mellanby, *Cambridge*

Dr H.G. Miller, *Macaulay Institute for Soil Research, Aberdeen*

B. Mould, *Department of Energy, London*

Dr Clifford Price, *Historic Buildings and Monuments Commission, London*

Dr P. Roberts, *International Flame Research Foundation, Ijmuiden*

Dr J. Skea, *Science Policy Research Unit, University of Sussex*

Dr R. Skeffington, *Central Electricity Research Laboratories, Leatherhead*

Dr F.B. Smith, *Meteorological Office, Bracknell*

Dr J.H. Stoner, *Welsh Water Authority, Haverfordwest*

Dr M.H. Unsworth, *Institute of Terrestrial Ecology, Penicuik*

Dr S. Warren, *Water Research Centre, Medmenham*

P.F. Weatherill, *British Gas Corporation, London*

Dr J.H. Weaving, *Solihull*

Dr M.L. Williams, *Warren Spring Laboratory, Stevenage*

M. Woodfield, *Warren Spring Laboratory, Stevenage*

Air Pollution, Acid Rain and the Environment

Edited by

KENNETH MELLANBY

CBE, ScD (Cantab.), FIBiol
Chairman of a Working Group appointed by
The Watt Committee on Energy

Report Number 18

Published on behalf of
THE WATT COMMITTEE ON ENERGY
by
ELSEVIER APPLIED SCIENCE PUBLISHERS
LONDON and NEW YORK

ELSEVIER SCIENCE PUBLISHERS LTD
Crown House, Linton Road, Barking, Essex IG11 8JU, England

Sole Distributor in the USA and Canada
ELSEVIER SCIENCE PUBLISHING CO., INC.
52 Vanderbilt Avenue, New York, NY 10017, USA

WITH 26 TABLES AND 64 ILLUSTRATIONS

© 1988 THE WATT COMMITTEE ON ENERGY
Savoy Hill House, Savoy Hill, London WC2R 0BU

British Library Cataloguing in Publication Data

Air pollution, acid rain and the environment.
  1. Environment. Pollution by acid rain.
  I. Mellanby, Kenneth, *1908–*   II. Watt
Committee on Energy   III. Series
363.7'386

  ISBN 1-85166-222-7

Library of Congress Cataloging-in-Publication Data

Air pollution, acid rain, and the environment/edited by Kenneth
Mellanby.
       p.   cm. — (Watt Committee report; no. 18)
    Bibliography: p.
    Includes index.
    ISBN 1-85166-222-7
    1. Acid deposition—Environmental aspects—Great Britain.   2.Air—
Pollution—Environmental aspects—Great Britain.   3. Sulphur
dioxide—Environmental aspects—Great Britain.   4. Environmental
protection—Great Britain.   I. Mellanby, Kenneth.   II. Series.
TD196.A25A366 1988
363.7'386'0941—dc 19   88-7178
                                                                                                        CIP

Printed in Great Britain by The Alden Press, Oxford

Foreword

The Watt Committee on Energy became active in the study of Acid Rain during 1982. Perhaps the only aspect of the subject that has become more certain during the subsequent five years is that the expression 'Acid Rain' is used loosely in public debate for a complex of industrial and environmental phenomena. Among these, Acid Rain in the straightforward meaning of the words—rain and perhaps snow having a significantly high level of acidity—is of only limited importance. To represent this perspective, therefore, the Watt Committee Executive decided that the study leading to the present Report should be entitled 'Air Pollution, Acid Rain and the Environment'.

The Watt Committee's interest in Acid Rain arises from the fact that, among its causes, the man-made ones arise from energy generation and use. The culpability of power station emissions remains a matter of dispute; stricter standards for the cleaning of emissions have recently been adopted in Britain. A key role may be played by ozone levels, increases in which seem to be affected by aerosols and by the design and efficiency of motor vehicle engines. Whatever the sources, the damage done by emissions depends very largely on climatic factors; geological factors are also important in some areas. Because it represents all the relevant disciplines, the Watt Committee is well placed to consider the relationship between the natural and the man-made factors and the effects of possible remedial strategies.

The effects of Air Pollution in its various forms —rain and snow, particulates, mists and so on—are widespread, and sometimes very noticeable, but scientific examination shows that popular accounts of them are liable to be exaggerated. In many advanced countries, including the United Kingdom, levels of atmospheric pollution over large areas are certainly lower than a century ago. Some of the effects are so long-term that they cannot now be reversed by remedial measures. This is true of some of the effects on buildings, for instance. Proposals for action should therefore concentrate on measures that promise a real improvement as a result of expenditure.

The Watt Committee's study of this subject has been in two phases. The first dealt with the nature of the problem, and culminated in the publication of Watt Committee Report No. 14 in 1984. That Report was divided into four sections, each of which was prepared by a sub-group of the working group: they dealt respectively with the fate of airborne pollution, vegetation and soils, fresh water and remedial strategy. In the second phase, these sub-groups have brought their sections up-to-date and a fifth sub-group was appointed to study buildings and non-living materials. The Watt Committee on Energy is grateful to the Chairman of the working group, Professor Kenneth Mellanby, and his team of volunteers, some of whom have had to make quite unreasonable inroads into their working-time in order to make their contributions to this Report.

The Watt Committee working group on Acid Rain had almost completed this Report when the Government announced its proposals for the privatisation of the electricity supply industry in the United Kingdom. As this Report goes to press, it is too early to predict their impact on environmental issues, but it is noteworthy that the new generating companies' responsibility for the environment does not yet seem to figure in the case as presented by spokesmen for Government and the industry. It is apparent that there will be commercial pressures on the companies to concentrate on development work that would bring early improvement in their financial performance, and it remains to be seen how the Government intends that the environmental issues be dealt with.

The conclusions reached by the working group appear in their place and I will not anticipate them. My role is to emphasise that the objective of the

Watt Committee is to improve the quality of public debate. To complex questions, such as Acid Rain, simple answers are unlikely to be reliable. It will be a real achievement if the reader of our Report on this emotive subject has a clearer factually based understanding of the options that are open for national and international decision-making.

G.K.C. PARDOE
Chairman, The Watt Committee on Energy

Contents

Introduction

The Watt Committee on Energy first set up its working group on Acid Rain in 1982, and the working group's first report (Report No. 14 of the Watt Committee) appeared in 1984. This is the second report of the working group, in which we attempt to give an up-to-date account of the effects of air pollution on our environment, particularly in the United Kingdom. We have given this Report the new title *Air Pollution, Acid Rain and the Environment*, in recognition of the fact that the problem is much wider than the expression 'acid rain' alone suggests.

It is clear that there is still public concern about damage by pollution to freshwater ecosystems, growing crops, trees, animals, building and metals. Those of us in this working group concerned with freshwater ecosystems consider that here the evidence is sufficient to show that man-made emissions, particularly of sulphur dioxide, are the most important cause of acidification and of the elimination of fish and other organisms from many lakes and rivers. This had been identified as a serious problem in several areas in Britain. For freshwater, therefore, there is good reason to believe that a reduction of emissions of sulphur dioxide would be beneficial, both in Britain and Scandinavia, though we still cannot foretell just how any reduction in emissions would affect deposition rates at a distance, nor can we be certain of the effect of any reduction in deposition on the acidity of particular lakes and rivers, and on their fauna and flora. Nevertheless, as evidence by the recent announcement of a limited flue-gas desulphurisation (FGD) retrofit programme for British power stations, there is now general agreement that it is desirable to maintain a downward trend in sulphur dioxide emissions.

With regard to trees and forests, the situation is quite different. Surveys in Britain show that today, as in the past, trees may be affected by climatic factors including drought and cold winters, by pests and diseases, and by pollution. In our woods it is always possible to find some unhealthy trees, and to observe premature leaf-fall, particularly after a dry spell. But we can find no evidence that there is serious damage to our woods and forests today that can be clearly attributed to air pollution or 'acid rain'. All the evidence available suggests that our woods are 'normal'. There are, for instance, cases of damage by drought which have been observed in one year, where the trees appear healthy in the next. However, we agree that continued monitoring to look for pollution damage, if damage does occur, is important.

The situation is very different from that which occurred in the Pennines in the past, when conifers died from what was clearly the effect of sulphur dioxide, and when many species of trees which flourish today could not be grown in our cities. Less welcome is the return of fungus diseases like black spot on roses and tar spot on sycamores to urban parks; these diseases were previously kept in check by high levels of air pollution.

We cannot confirm some alarming reports which have appeared in the press and on television concerning widespread damage to trees in Britain. Reports from other countries also seem often to have been exaggerated. Serious damage to trees has occurred in several European countries, particularly in the German Black Forest and in Switzerland (and, though perhaps not as well known, most strikingly in Czechoslovakia), and this has been studied by scientists. However, several observant naturalists confirm my own observation that for the most part the beauty of European forests is largely unimpaired, though serious local damage does exist and can be found when looked for.

When we were preparing our previous report, it was generally accepted that, with the reduction of ground-level concentrations of sulphur dioxide in our urban areas, damage to buildings was no longer a serious problem. However, various reports, including that of the Environment Committee of the House of Commons (1984), have suggested that damage to cathedrals and other historic buildings is still serious and may even be increasing. This problem has been studied by a new sub-group, whose findings are included in this Report. The problem is clearly complicated. With living organisms, damage from pollution is generally recognizable at the time of exposure or soon after. With buildings, nothing may appear to be happening for many years, and then, perhaps when pollution is no longer at a high level, stone surfaces may begin to disintegrate. While it is clear that high levels of sulphur dioxide are a serious cause of damage, and that their reduction has been beneficial, we cannot yet be certain whether or not oxides of nitrogen cause comparable damage. This is clearly a subject where more research is needed.

Although there may still be uncertainties regarding the effects of different levels of air pollution on our environment, we believe that we should evaluate the various possible methods of reducing emissions and ameliorating their effects. In this Report the section on 'remedial strategies' deals with this subject.

It is important that, if emissions are to be controlled, the most effective techniques should be available. Also the beneficial effects must outweigh any harmful side effects. Thus, even though the limestone requirement for FGD at one 2000 MW(e) station is a fraction of one per cent of the national limestone production, concern has been expressed that the specific quarries used should not be ones where important features of the landscape might be destroyed. A possible requirement to dispose of some of the product material to landfill sites and the treatment of other effluents are also questions of some environmental concern. Environmental pressure groups most concerned with the pollution may find themselves also concerned with minimising the impact of the steps taken to control it.

Finally, speaking personally, and not wishing necessarily to commit my colleagues on the working group, or the Watt Committee itself, I am reasonably optimistic about the future. I believe that when we understand the problems more completely, we will take the appropriate steps to solve them. Also I think that it is probable that many of the forecasts of doom and destruction being made by some 'environmentalists' will prove as inaccurate as those made on other environmental topics 15 or 20 years ago. We were told in 1969 by Professor Paul Ehrlich that before 1980 the world's oceans would be dead, and that the Chinese and Japanese would suffer starvation from the disappearance of fish and other sea food from their diet. In 1972 the publication 'Blueprint for Survival' forecast that global supplies of silver, gold, mercury, lead and zinc would be exhausted, or at least nearing exhaustion, by 1987. It also said that food shortages would occur as Britain's farms would soon suffer substantial decreases in productivity—when in fact in the next ten years there was the greatest increase in yields ever known in any ten-year period in our history. When, at the time, I queried these gloomy forecasts, I was accused of 'complacency'. I shall be happy to be so denigrated again today. I believe that, if we behave reasonably sensibly, our descendants will still see the beautiful forests and productive rivers and lakes that have existed for so many hundreds of years, and that their quality will, in time, be improved and not destroyed. Nevertheless, pressure for improvement must be kept up, but this must be based on facts and not fiction. Without this pressure, there is always the risk that the relentless drive for economy and profit, with the corner-cutting to which this often leads, could result in future losses outweighing the gains that are clearly obtainable.

KENNETH MELLANBY
Cambridge

References

Anon. (1972). A blueprint for survival. *The Ecologist*, **2**, 1–43.
EHRLICH, P. (1969). Eco-catastrophe. *Ramparts*, 24–8.

Section 1

Production and Deposition of Airborne Pollution

Barry Smith

Deputy Chief Scientific Officer, Boundary Layer Research Branch, Meteorological Office, Bracknell, Berks

This paper presents the work of a sub-group of the
Watt Committee working group on Air Pollution,
Acid Rain and the Environment.

Membership of Sub-group

Dr F.B. Smith (Chairman)

Dr H. ApSimon
Dr A. Cocks
Dr G. Dollard
Dr B. Fisher
A. S. Kallend
M. I. Manning
A. R. Marsh
Dr P. Roberts
Dr M. H. Unsworth

1.1 INTRODUCTION

In the first report on 'Acid Rain' (Watt Committee, 1984), the general processes of emission, transport, transformation and deposition of air pollution were described (see Fig. 1.1). The importance of dispersion by atmospheric turbulence was stressed since it is this dispersion which enables the plume to mix with surrounding 'clean' air carrying oxidants which can transform some of the major primary pollutants through a series of complex reactions into secondary pollutants whose capability for ecological damage may be radically different from that of the original primary pollutants. The dispersion may also bring an elevated plume down to ground where some of the pollutants may undergo gradual deposition as a result of sedimentation, impaction or chemical adsorption—a process called 'dry deposition'. The rate of dry deposition depends on the airborne concentration C and the state and nature of the surface and overlying atmosphere; in practice it is usually assumed equal to the product of C and a so-called deposition velocity v_d. For aerosols and most gases v_d is usually small and at most is only a few centimetres per second. However, for some very reactive gaseous species, such as nitric acid, the deposition velocity may considerably exceed this in windy conditions over tall vegetation.

When driven by the wind, cloud or fog drops may be deposited onto vegetation elements with high efficiency, giving rise to 'occult deposition' of pollutants contained within the drops.

Dispersion is also partly responsible for drawing pollution into precipitating cloud where through one process or another (see the first report) it may be removed: a mechanism called 'wet deposition'. The removal rate, when effective, is rather rapid since rain is a relatively efficient removal mechanism, but in general it is only operative for some small percentage of the time, typically less than 10%. Roughly three-quarters of the total European sulphur emission is redeposited on European countries, according to the European Monitoring and Evaluation Programme (EMEP) Routine model: the rest is either deposited into the sea or is carried out of the area by the wind. The new EMEP model reduces this fraction to some 60% for the year 1980. Due to known biases in this model it is likely the true figure lies between these estimates. Over Europe as a whole, dry deposition has a magnitude about twice that of the wet deposition, although in some particularly wet areas, and in very remote areas, the wet deposition exceeds the dry.

Since the first report, progress has been made in a number of areas:

(i)   UK emissions from vehicles are now estimated in a more realistic manner, so that significant changes have had to be made to previous estimates for NO_x and carbon monoxide.

(ii) Estimates of UK emissions for a number of pollutants are now available for an extra four years.

(iii) Trends in emissions and depositions are being examined for simple relationships, although strictly many more years of good quality data will be required to demonstrate a real connection unequivocally.

(iv) The Great Dun Fell Project, in Cumbria, designed to study the fate of SO_2 in cloud, variations in chemistry with altitude and deposition processes, is now producing very useful data. Four organisations (the Institute of Terrestrial Ecology (ITE), the University of Manchester Institute of Science and Technology (UMIST), The Harwell Laboratory of the UK Atomic Energy Authority (UKAEA) and the University of East Anglia (UEA)) are collaborating in this project. Measurements show an increase in the concentrations of most ions with altitude which, in conjunction with an increase in precipitation, implies a markedly larger wet deposition at the top of the fell than at the bottom. The physio-chemical processes involved in cloud formation as air rises over the fell, and of the seeder-feeder processes in precipitation intensification, are being investigated by the development of models and comparison with field measurements (Fowler *et al.*, in press; Choularton *et al.*, in press). In the 'captive' cloud system studied, the process of dry-air entrainment under non-adiabatic

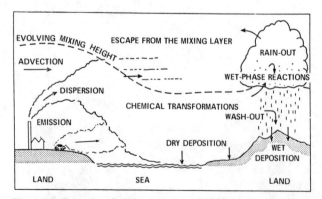

Fig. 1.1.   Processes involved in the deposition of atmospheric pollutants.

conditions has been identified as an important source of additional oxidant that would be of equal importance in 'free' cloud systems.

Measurements of simultaneous changes in hydrogen peroxide and ambient sulphur dioxide levels have yielded important information on SO_2 oxidation rates within clouds in general.

(v) The Central Electricity Research Laboratories, Leatherhead, have also been carrying out important measurements at Great Dun Fell. Their aim has been to understand the fate of sulphur dioxide drawn into clouds in a rather controlled environment. This has been achieved by releasing a plume of sulphur dioxide into an airstream ascending the fellside when cloud forms within it before reaching the crest, and studying the relative concentrations of sulphur dioxide and hydrogen peroxide. Results broadly confirm laboratory expectations concerning the rate at which the hydrogen peroxide controls the uptake and oxidation of the SO_2.

(vi) Several other important projects are also now underway. The Central Electricity Generating Board (CEGB) have started a collaborative programme on Photochemical Oxidants to be carried out in collaboration with several British Universities including UEA, UMIST, Cranfield and Cambridge at a cost of more than £1 million over three years. There has also been recent work on natural emissions including that of UEA (and others) on sulphur emissions from plankton blooms in the North Sea and elsewhere.

(vii) Complex models are now under active development in Europe to study episodes and the formation of photo-oxidants in greater detail than has previously been possible.

(viii) The Chief Scientist's Group of the Energy Technology Support Unit (ETSU) at the Harwell Laboratory of the UKAEA have recently produced a report (Derwent, 1986) on work on the long-range transport of nitrogen species using a simplified version of the EMEP (MSC-W) model designed for sulphur species. Using emission fields for NO for the UK and Europe, wet, dry and total deposition fields have been achieved appropriate to meteorological conditions extending from February 1981 to October 1982. An extract from the report is given in Fig. 1.2, which shows a map of the model's implied annual total deposition

over the UK of NO_x and its derivatives from all known UK and European sources.

(ix) Significant progress has now been made, by complex statistical modelling, towards answering the question posed in the first report, namely: 'Is the long-term average deposition of sulphur species in a sensitive region of Europe resulting from any known source linearly proportional to its emission magnitude, within acceptable limits?'

(x) Since the first report advances have been made in the area of background NO_x and ozone monitoring. The tenth report of the Royal Commission on Environmental Pollution (1984) and the report of the House of Commons Select Committee on the Environment (1984) both highlighted the paucity of monitoring of these pollutants, and the government in its response accepted the need

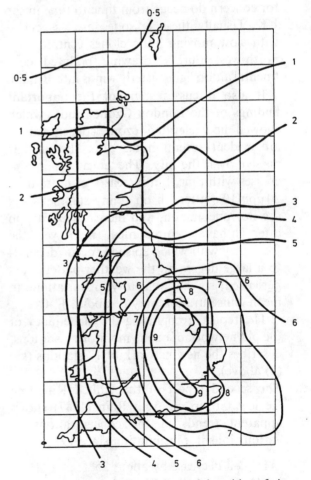

Fig. 1.2. Model prediction of the total deposition of nitrogen oxides and their derivatives (in $kg\,N\,ha^{-1}\,yr^{-1}$) from UK and other European sources (Derwent, 1986). Experience with similar modelling studies for sulphur species suggests that some of the modelling assumptions may somewhat distort the field and that real depositions may be less peaked in the East Midlands with larger values than shown in the remoter mountainous areas of high rainfall.

to expand the network. The Department of the Environment (DoE) asked Warren Spring Laboratory (WSL) to make proposals for an enlarged network and these were discussed at at expert meeting in March 1985. The recommendation from that meeting was for a national network of about 17 stations over the UK which is planned to be completed in 1988.

(xi) In 1985 the DoE set up the Photochemical Oxidants Review Group (similar in nature to the Review Group on Acid Rain, composed of experts from government laboratories, industry and universities) to address the photo-oxidants issue and prepare a report summarising UK ozone data up to the present. The report, which has recently been published (United Kingdom Photochemical Oxidants Review Group, 1987), describes how episodes above certain prescribed thresholds for concern do occur from time to time in the UK. Usually these episodes are associated with slow moving anticyclones centred over north-west Europe drawing polluted continental air on light easterly winds into Britain.

It also discusses some of the important findings of the London Ozone Study, which showed how elevated ozone concentrations are evident several hours after the air has passed over the city. The enhancement increases with time, but increases of more than $20\,\mathrm{ppb}(10^9)$ after 3 h on sunny summer days are not uncommon, and increases of 70 ppb after 7 h have been recorded. In any year the frequency with which episodes occur depends to a large degree on the weather patterns experienced, so that year-to-year variations in mean concentration may be about $\pm 40\%$.

The report, recognising the inadequacy of the old network of ozone monitoring stations, describes the new network of 17 stations (see (x) above).

(xii) The second report of the United Kingdom Review Group on Acid Rain (1987) shows apparent trends in the concentrations of various ions in precipitation since 1978:

H^+ —a decrease by about 25%
SO_4^{2-} —a smaller decrease
NO_3^- —little change
NO_3^-/SO_4^{2-} —clear increase in Scotland, less clear increase in England (overall increase by a factor of 3 in the last 30 years)

H^+/SO_4^{2-} —marked decrease
H^+/NO_3^- —small decrease

The contribution from UK emissions is a rather small fraction of the whole sulphate at the rather remote sites in the monitoring network. This fraction appears to have decreased more or less in parallel with decreasing UK emissions, but this conclusion is rather uncertain due to the large variability in depositions from year to year.

Two interesting recommendations in the report are: (a) effort should be placed in developing improved ionic-deposition samplers for snow; and (b) a small network should be established (linked to precipitation sampling) to measure nitric acid/nitrate aerosol, ammonia/ammonium aerosol and sulphate aerosol.

(xiii) Recently increasing attention has been given to the probable role of ammonia in acidification, particularly downwind of areas of concentrated agriculture. The presence of ammonia can enhance the potential uptake and oxidation of sulphur dioxide in cloud droplets and rain, and hence alter the spatial footprint of deposition. Subsequent deposition of some ammonium compounds may also lead to a liberation of H^+ ions and additional soil acidification through microbial processes of nitrification.

1.2 EMISSIONS FROM VEHICLES: REVISED PROCEDURE

1.2.1 The old method

In years prior to 1983, Warren Spring Laboratory estimated the emission from petrol- and diesel-engined vehicles on the basis of a fuel consumption multiplied by an emission factor, so that, for example, the UK annual emission of NO_x from petrol-engined vehicles was calculated as:

$$Q_{NO_x} = F_{P,NO_x} C_p$$

where F_{P,NO_x} is the emission factor appropriate to NO_x from petrol-engined vehicles and C_p is the national annual consumption of petrol. A similar expression was used for DERV, and also for other pollutants.

The petrol factors used were based on a study of 201 cars done by the Motor Industries Research Association (MIRA)(Williams & Southall, 1976). This study used 'as received' cars and the 201 car sample was designed to be representative of the UK petrol-engined fleet. However, this work was old —the Report dealt with a fleet typical of around 1975, and the diesel factors were based on American work from even earlier (around 1970–1974).

The petrol emission factors were obtained from the MIRA Study which measured emissions from the Standard Test Cycle using a dynamometer (ECE Regulation 15) which is typical of urban driving in which the vehicle would have an average speed of about $19 \, km \, h^{-1}$. The diesel factors were based on studies on road-routes typical of urban/suburban drives.

1.2.2 The new method

The new method is based on a device called a mini-constant volume sampler, built at Warren Spring Laboratory, which enables on-the-road emissions to be measured, rather than relying on dynamometer results. An initial test programme showed that curves of emissions against speed could be generated and that emissions, particularly of CO and hydrocarbons, were strongly dependent on speed. A representative sub-sample totalling 20 cars, of a later 1982/83 MIRA 204-car study was taken (Williams & Everett, 1983) and full emission–speed curves were generated (see Fig. 1.3a–c). These curves are now used as the basis of the new method for UK emissions from motor vehicles from 1983 onwards. They could also provide a reasonable approximation to emissions from 1980 to 1982 since the 204 (20) car samples were representative of the 1981/82 car fleet. Systematic re-updating will be required in the future.

In order to use these curves, it is necessary to know the vehicle-kilometres travelled (i.e. the 'traffic activity') as a function of speed over the UK,

and this information has been obtained by WSL from traffic survey data.

The old and new methods give quite different results for CO, less so for NO_x, and fortuitously almost the same for total hydrocarbons (see Table 1.1) in the 1983 Watt Committee Report.

1.2.3 Accuracy of national emissions

The range of results of the measurements of vehicle emissions has important implications for the accuracy and precision of the estimates of the national emissions. Taking the midpoint of the envelope of the $NO_x/CO/HC$ curves produced the values in Table 1.1. If, however, we take the upper and lower estimates then the errors on the petrol-vehicle emissions are for $NO_x \pm 54\%$, $CO \pm 74\%$ and $HC \pm 43\%$. These may be pessimistic since one could argue that the envelope may represent three standard deviations, but on the other hand the subsample only had 20 cars in it. Overall then, taking these uncertainties together with those associated with other sources (based on very sparse

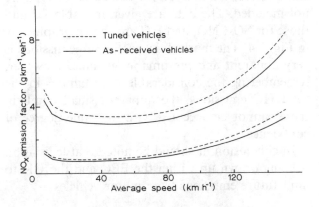

Fig. 1.3b. NO_x (as NO_2) emission factor (envelope of measured data): variation with speed.

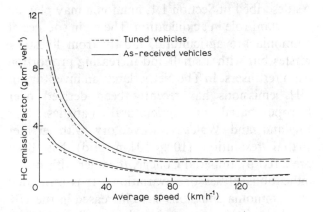

Fig. 1.3c. HC emission factor (envelope of measured data): variation with speed.

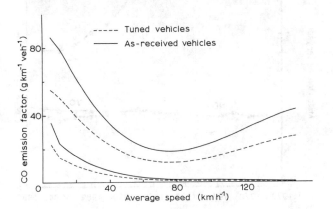

Fig. 1.3a. CO emission factor (envelope of measured data): variation with speed.

Table 1.1 UK 1983 emissions (10^3 t year^{-1})

	NO$_x$ (as NO$_2$)		CO		HC	
	Old	New	Old	New	Old	New
Vehicles (petrol & Derv)	508	688	8107	4437	538	531
Δ%		+35%		−45%		−1%
Total UK uncertainty	1639	1819 ±45%	8939	5269 ±65%	1640	1633 ±50%
Δ%		+11%		−41%		−0%

data), the errors in the total *national* emissions (incorporating all sources) lead to the very tentative estimates given in Table 1.1. The assessment of the magnitude of these errors will be improved as more detailed data become available.

1.2.4 Update of SO$_2$ and NO$_x$ emissions for the UK

An update is given which incorporates: (i) the new motor vehicle calculations; and (ii) data for four additional years, namely 1983 to 1986. Revised emissions for CO and HC for years prior to 1975 are not included. The data are given in Table 1.2 and those for SO$_2$, NO$_x$ and HC are shown graphically in Fig. 1.4. The marked decline in SO$_2$ emissions is very apparent and presumably genuine. However, remembering the considerable uncertainties in the total HC emissions, the apparent small climb in these cannot be accepted with the same degree of confidence.

In conclusion, it should be noted that legislation is being drawn up within the EEC and the UK to limit future emissions from motor vehicles.

1.3 SOURCES OF AMMONIA

As described in Section 1.1, ammonia may play an important role in acidification. The main sources of ammonia are agricultural, mostly from livestock wastes but with a small and increasing proportion from fertilisers. In The Netherlands an inventory of NH$_3$ emissions has recently been derived over Europe based on agricultural statistics. For England and Wales an inventory with greater spatial resolution (10 × 10 km grid) has been prepared by Imperial College, together with broad indications of seasonal variations. It is estimated that ammonia emissions have increased in the UK by about 50% since 1950 with a similar trend on average over the rest of Europe.

The density of NH$_3$ emissions varies considerably geographically. Over England and Wales estimated emissions are lower in the arable farming areas to the East (< 1 t km^{-2} year^{-1}) than over the western part of the country with more livestock (up to 5 t km^{-2} year^{-1}).

Table 1.3 shows estimated annual emissions for England and Wales from different livestock sources giving a total of 300 kilotonnes of NH$_3$ per year. On a similar basis a figure of 70 kilotonnes per year of NH$_3$ has been derived for Scotland. Monitoring of ambient NH$_3$ and NH$_4^+$ levels is also in progress, for example at the University of Essex, and further measurements are planned.

1.4 TRENDS IN EMISSION AND DEPOSITION

The Report of the United Kingdom Review Group on Acid Rain (1983) prepared for the DoE, was mentioned briefly in the last Report. The Group was, at that time, unable to provide a complete picture of wet deposition over the United Kingdom

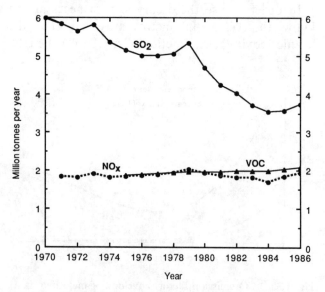

Fig. 1.4. UK emissions since 1970 (million tonnes per year).

Table 1.2 UK emissions of SO_2, NO_x (as NO_2), CO and volatile organic compounds (VOCs), since 1971 (million tonnes per year)

	1971	1972	1973	1974	1975	1976	1977	1978	1979	1980	1981	1982	1983	1984	1985	1986
SO_2 Power stations	2·80	2·87	3·02	2·78	2·82	2·69	2·75	2·82	3·10	2·87	2·71	2·62	2·54	2·50	2·51	2·60
All sources	5·83	5·64	5·80	5·35	5·13	4·99	4·99	5·04	5·33	4·68	4·23	4·02	3·71	3·53	3·56	3·74
NO_x Power stations	0·76	0·73	0·81	0·73	0·77	0·77	0·80	0·81	0·88	0·85	0·82	0·77	0·76	0·62	0·73	0·78
Vehicles[a]	0·54	0·57	0·60	0·59	0·58	0·60	0·62	0·65	0·66	0·67	0·65	0·67	0·69	0·72	0·74	0·78
All sources	1·85	1·82	1·95	1·82	1·83	1·86	1·90	1·93	2·02	1·93	1·86	1·81	1·82	1·69	1·84	1·94
CO All sources[a]	—	—	—	—	4·76	4·86	4·98	5·14	5·23	5·21	5·07	5·28	5·27	5·19	5·37	5·60
VOCs All sources	—	—	—	—	1·86	1·89	1·91	1·94	1·97	1·97	1·95	1·98	1·98	1·98	2·02	2·07

[a] Principally (80%) from vehicles.

Table 1.3 Total NH₃ emissions in England and Wales

	Tonnes per year	Percentage
Cattle	181 711	60·0%
Sheep	62 118	20·5%
Poultry	25 408	8·4%
Pigs	19 277	6·4%
Fertiliser	12 357	4·1%
Horses	1 666	0·6%
Total	302 537	100·0%

due to incomplete coverage of the land area by monitoring stations observing sufficiently good scientific protocol. They also concluded that the limitations of past data also precluded the identification of detailed relationships between emissions and pollutant concentrations in rain, although a fairly clear trend of increasing nitrate concentration was discernible over a period of increasing NO_x emissions.

Following their recommendations, networks have been established to give a much more comprehensive coverage over the UK. At ten primary stations daily measurements are made of concentrations of the ions pH/H^+, NH_4^+, Na^+, K^+, Ca^{2+}, Mg^{2+}, NO_3^-, Cl^-, SO_4^{2-} and PO_4^{3-} in precipitation using wet-only collectors as well as of airborne SO_2, NO_2 and particulate sulphate. Sixty secondary stations have a more limited objective, all making weekly measurements of at least pH/H^+, NH_4^+, Na^+, NO_3^-, Cl^- and SO_4^{2-} in rain using bulk collectors, but many analysing for all the measured ions in the primary network (see Fig. 1.5). The secondary network became operational in January 1986.

Considering the measurements currently available from both the UK and other networks, trends in concentration or of deposition with time sometimes appear to be evident. However, they need to be viewed with caution. For example, systematic bias of meteorological patterns is well-recognised, and, over a time-scale which may be as long as decades, this factor may be dominant. This effect can be seen most clearly, for example, in the European Air Chemistry Network data which show trends over a 25-year period which differ widely for different regions of Europe. There has also been progress by Davies *et al.* (1988) at the University of East Anglia, in establishing that changes in atmospheric circulation do make a contribution to variations in acidic deposition on a time-scale of years to decades. This kind of influence inevitably means that records over even longer periods of time are required to produce statistically significant relation-

ships between emissions and concentrations. This caveat is very familiar to those responsible for interpreting measurement data, but who are often required to give indications of possible relationships —albeit without satisfactory significance—in response to expressed concerns about the environment.

Measurements made by the Institute of Terrestrial Ecology in Northern Britain over the period 1979–83 further illustrate the problem of interpretation. As we have seen earlier, UK emissions of SO_2 decreased by some 26% over that period. However, the concentrations and rates of deposition of non-marine sulphate show no systematic trend, increasing at some sites and decreasing at others, again indicating that the relationship between emissions and deposition is a complex one. This is confirmed by the change in the ratio of deposited nitrate to non-marine sulphate from 0·3 in 1980 to 0·5 in 1983 and by the 40% increase of weighted mean concentrations of the sea-salt derived ions sodium and chloride over the same period. Deposition of acidity at the ITE sites did

Fig. 1.5. A map of the Acid Deposition Monitoring Network primary (◇) and secondary (▲) sites. All primary sites will have secondary-site bulk-collectors to provide inter-comparison.

show a decrease over the period but clearly this cannot have been associated with changes in sulphate. Moreover, since the ratio of nitrate to sulphate actually increased, it cannot have been associated primarily with changes in nitrate either, and must therefore be related to the concentrations of other ions, notably cations such as ammonium.

Outside the UK in Scandinavia, measurements at Birkenes in south Norway of non-marine sulphate in precipitation from air masses which had traversed the UK, made during the period 1977–82 within the scope of the EMEP programme, appeared to follow the trends in UK emissions of SO_2 over the same period. However, the statistical significance of the relationship was low and the results for 1980 were anomalous for unexplained reasons. Taking these results, together with those from the UK, the only wholly safe conclusion which may be drawn is that a need exists for longer sequences of reliable data and improved methods of data analysis. Thus although progress is being made, our conclusion about what presently available data reveal is not materially different from that of the previous report (Watt Committee, 1984) or from that of the first report of the Review Group on Acid Rain (1983).

1.5 MODELLING: THE DEVELOPMENT OF A HIERARCHY OF MODELS

In the first report three main types of model were briefly discussed: the statistical models, the operational models and the complex models. The degree of simplification is greatest in the statistical, and least in the complex, models (although far from eliminated even in these) and each type has its own virtues and weaknesses as well as its own particular area of application. At that time, all European models were of the first two types. Now the situation has begun to change. More complex computer models are currently under development to simulate the dispersion, transformation and deposition of atmospheric pollutants over long distances. Few of these are yet fully operational. They involve following the history of pollutants emitted as though transported through a three-dimensional assembly of grid cells with winds varying in space and time. Detailed chemical models are included to describe the large range of chemical reactions as pollutants are transported and mixed. So extensive are the calculations involved that, using large computers such as the Cray 1, the computer time required is 20–25% of the real time simulated. Considerable

numerical problems arise (in for example avoiding artificial 'numerical diffusion' resulting from the numerical integrations), and in spite of their sophistication these models are still limited by such factors as the accuracy of the windfield deduced from available observations, and the detailed character of precipitation.

Great effort is also involved in compiling the necessary data including emission inventories for many species of pollutants (e.g. SO_x, NO_x, various hydrocarbons, NH_3, etc.) with a detailed grid resolution plus specification of major individual sources.

1.5.1 Modelling studies at the Harwell Laboratories

Derwent (UK Photochemical Oxidants Review Group, 1987) has developed a sophisticated photochemical model, albeit with a fairly simple meteorological transport scheme, which has produced several interesting conclusions: the concentration of ozone close to the ground is strongly influenced by the diurnal variations in the dry deposition velocity; hydrocarbons and oxides of nitrogen have completely different roles in ozone formation; the highest peaks in ozone concentration require persistent slow-moving anticyclones; aromatic hydrocarbons are of dominant importance in the early build-up of ozone and PAN (peroxyacetyl nitrate). The sources of these hydrocarbons in the UK are, in order of importance, solvents, petrol exhausts, natural gas leakages, petroleum refineries, industry and evaporation of petrol.

1.5.2 The PHOXA project

The PHOXA project arose from a collaborative research programme between The Netherlands and the Federal Republic of Germany, and involves the use of such a complex model—the SAI model. The Organisation for European Cooperation and Development (OECD) and the European Economic Community (EEC) are now also collaborating in this project, with OECD providing an emissions inventory. The aim is to study carefully selected episodes of both deposition of acidifying substances and photochemical oxidation. Current plans include the following episodes:

25–28 February 1982:   an acid deposition episode

24–26 July 1980:
29 May–2 June 1982:   } Three photo-oxidant episodes
3–6 June 1982:

The photo-oxidant episodes were chosen on the basis of anticyclonic situations with observations of high ozone concentrations (above 80 ppb (10^9))resulting from photo-oxidation, but with little contribution from Eastern Europe since detailed emission data are not available there. As well as using the OECD emissions inventory, calculations will also be performed simulating changes in emissions of NO_x and volatile hydrocarbons to indicate the potential effectiveness of possible control measures in reducing ozone concentrations in these particular meteorological situations. Because of the extensive computer demands only a very limited number of investigations are possible. Results to date are in many ways encouraging but sometimes still show significantly large discrepancies between predicted and observed ozone concentration levels.

1.5.3 Proportionality

One of the questions of vital importance in the search for optimum ways of reducing environmental damage is: 'if the emission of a pollutant is reduced, will an approximately proportional reduction in the related environmental damage occur?'. Unless the answer is 'yes', at least on some meaningful time-scale, then there can be little incentive to embark on a costly programme to reduce emissions. Of course, if the pollutant could be eliminated entirely, the related damage must ultimately cease, but such elimination is usually impossible to achieve. A substantial reduction is all that can be realistically hoped for.

The question really hides two questions: (1) 'if emissions are reduced will related depositions be reduced in approximate proportion?'; and (2) 'if depositions are reduced will environmental damage be reduced correspondingly?'. Whilst the second question is outside the remit of this chapter, the probability is that in many situations the environmental response is likely to be noticeably non-proportional to changes in deposition.

The first question was briefly considered in the first report (Watt Committee, 1984). As discussed earlier, trends in time apparent in current data sets are largely insufficient to answer the question with any degree of certainty. Modelling is therefore the best approach for the time being. It is not until rather recently that statistical models have been developed at the Meteorological Office by Smith (1987) and at the Central Electricity Research Laboratories by Clark (1987) (and, since, elsewhere)

which include, in a highly parametrised but qualitatively effective fashion, the basic non-linear character of the wet-removal processes of a pollutant like sulphur dioxide and its oxidised forms. By expressing the fate of sulphur emissions in a great ensemble of possible situations in terms of probabilities, the related deposition fields averaged over a long period of time (strictly over several years) can be assessed, and the response to changes in these emissions determined. These models, although sufficiently simple to run on a micro computer, include most of the processes that affect the pollution, namely different source configurations and magnitudes, oxidation of sulphur dioxide to sulphate, dry deposition (with different deposition velocities for SO_2 and sulphate), plume growth by turbulence, the effect of additional sources on the way to the receptor, the passage in and out of 'wet regions' where precipitation occurs, passage over mountains and the effect these have on rainfall, and finally non-linear wet deposition with variable efficiency. The models show that the related total (wet plus dry) depositions downwind from a particular source are remarkably insensitive to these processes, except the magnitude of the intermediary sources, and that the response in the related total deposition to changes in the initial source strength is approximately proportional, provided the receptor is beyond a few hundred kilometres from the source (see Table 1.4). The criterion for describing the relationship as 'approximately proportional' is taken to be that a 50% reduction in source strength would produce a reduction in deposition in the range 40 to 60%.

Table 1.4 summarises some of the main conclusions. Dry deposition is almost always approximately proportional (the response is always slightly greater than expected on the basis of exact proportionality because of the indirect effect of the non-linearity in the wet deposition). The three source types referred to in the table are displayed schematically in Fig. 1.6, and it can be seen that they encompass the most probable configurations of sources. The model has been run with no other sources in between the initial source and the receptor (Table 1.4, results (a)), and also with a continuous array of sources equivalent to a highly industrialised region (Table 1.4, results (b)).

In the region where the wet deposition behaves non-proportionally the actual behaviour depends on the detailed mechanisms of oxidation and removal. Total (wet plus dry) deposition from a given source behaves much more proportionally than wet deposition, because apart from the near

Table 1.4 Model results showing the distance *d* beyond which deposition is 'approximately proportional' to emission

Source type	Intermediate[a] sources	Values of d^b for:	
		Wet deposition (km)	Total deposition (km)
Broad dispersed	No	300	250
sources	Yes (a)	150	50
	(b)	>2000	50
Single isolated	No	500	200
large point source	Yes (a)	Values not currently available:	
	(b)	probably similar to those for	
		the broad dispersed plume.	
Composite plume from	No	3000	0
cluster of large	Yes (a)	1400	0
sources	(b)	>2000	0

(Although the model is remarkably robust to changes in the values of the various input parameters, these values of *d* should be taken to indicate only in a broad sense values that should apply in reality.)

[a] (a) intermediary source-strengths unaltered when initial source-strength reduced; (b) initial and intermediary source-strengths reduced by same proportion.

[b] Distance from East Midlands source area to sensitive lakes of south Norway is about 800 km.

proportionality of dry deposition, those conditions leading to highly non-proportional wet deposition inevitably lead to a smaller contribution of wet deposition to the total deposition. It is not likely for wet deposition from a given source to be simul-

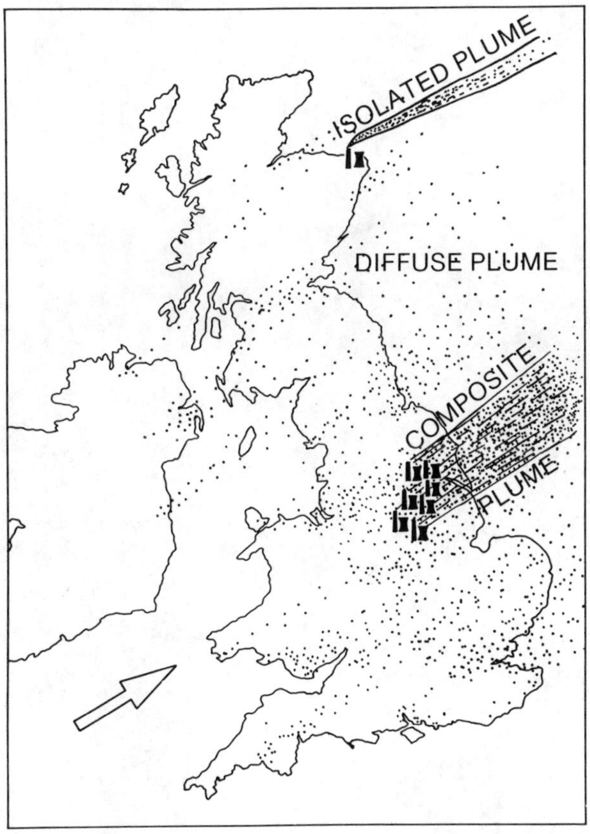

Fig. 1.6. The three types of sources considered in the 'proportionality' question.

taneously a dominant proportion of total deposition and also highly non-proportional to the emission strength.

Overall these modelling studies indicate that the model-prediction of the long-term average distribution of total and wet deposition over Europe would not be significantly improved by the assumption of non-proportional wet deposition, except in areas dominated by large local sources.

Nevertheless, these conclusions clearly do not negate the importance of research into non-linear air chemistry since there are many problems and situations of ecological importance for which the proportionality-assumption may not be valid.

1.6 EPISODES

Whilst, as indicated above, reductions in emissions will reduce long-term total depositions in approximate proportion, the same is unlikely to be true in episodes of concentration or deposition in rain or snow. Although the effect is at present unquantified, it is conceivable that equally serious episodes will still occur following a reduction in emissions, although perhaps rather less frequently. Thus if any aspect of the environment is prone to significant damage from short-lived episodes, this damage may still occur unless the reduction in emissions is very substantial indeed. These conclusions are, however, at present very tentative, and further work in this area is urgently required.

REFERENCES

CHOULARTON, T. W., GAY, M. J., FOWLER, D., CAPE, J. N. & LEITH, I. D. (in press). The influence of altitude on wet deposition composition between field measurements at Great Dun Fell and the predictions of the seeder–feeder model. *Atmos. Environ.*

CLARK, P. A. (1987). The influence of the nonlinear nature of wet scavenging on the proportionality of long term average sulphur deposition. In *Interregional Air Pollutant Transport: the Linearity Question*, ed. J. Alcamo, H. ApSimon & P. Builtjes. IIASA RR-87-20, Laxenburg, Austria.

DAVIES, T. D., FARMER, G. & BARTHELMIE, R. J. (1988). The contribution of atmospheric circulation variability to the changing pattern of 'acidic deposition' in Europe. Final Report of Dept. of Energy Contract PECD 7/10/101, available from the Library, Univ. of East Anglia, Norwich, UK.

DERWENT, R. G. (1986). The nitrogen budget for the UK and NW Europe, Energy Technology Support Unit Report 37, ETSU, AERE, Harwell, UK.

FOWLER, D., CAPE, J. N., LEITH, I. D., CHOULARTON, T. W., GAY, M. J. & JONES, A. (in press). The influence of altitude on rainfall composition. *Atmos. Environ.*

House of Commons Select Committee on the Environment (1984). *Acid Rain*, Session 1983–4, Vols I and II. HMSO, London (Sept. 1984).

Royal Commission on Environmental Pollution (1984). Tenth Report. Chairman: Sir Richard Southwood. HMSO, London.

SMITH, F. B. (1987). The response of long-term depositions to nonlinear processes inherent in the wet removal of airborne acidifying pollutants. In *Interregional Air Pollutant Transport: the Linearity Question*, ed. J. Alcamo, H. ApSimon & P. Builtjes. IIASA RR-87-20, Laxenburg, Austria.

United Kingdom Photochemical Oxidants Review Group (1987). Interim report: ozone in the United Kingdom. Dept of the Environment, South Ruislip, UK (Feb. 1987).

United Kingdom Review Group on Acid Rain (1983). Acid deposition in the United Kingdom. Warren Spring Laboratory, Stevenage, UK.

United Kingdom Review Group on Acid Rain (1987). Second report: acid deposition in the United Kingdom 1981–1985. Warren Spring Laboratory, Stevenage, UK.

Watt Committee (1984). *Report No. 14—Acid Rain*. The Watt Committee on Energy, London.

WILLIAMS, C. & SOUTHALL, M. (1976). Inservice emissions of cars manufactured to meet ECE Regulation 15. MIRA Report K 3404, Motor Industries Research Association, Nuneaton, UK.

WILLIAMS, C. & EVERETT, M. J. (1983). Inservice emissions of 204 vehicles manufactured between 1971–1982. MIRA Report K 32126, Motor Industries Research Association, Nuneaton, UK.

Section 2

Vegetation and Soils

William Binns

Formerly with the Forestry Commission, Farnham, Surrey

This paper presents the work of a sub-group of the
Watt Committee working group on Air Pollution,
Acid Rain and the Environment.

Membership of Sub-group

Dr W. O. Binns (Chairman)

Dr J. N. B. Bell
Dr M. J. Chadwick
Dr J. A. Lee
Prof. T. A. Mansfield
Prof. K. Mellanby
Prof. H. G. Miller
Dr R. A. Skeffington
Prof. M. H. Unsworth

The Watt Committee report (1984) concluded that SO_2, NO_2, O_3 and acid precipitation all had the potential to contribute to plant injury. It judged the most serious gaps in our knowledge to be the significance of ambient levels of acid precipitation when accompanied by gaseous pollutants, and the interactions between pollutants and the many natural stresses imposed by the environment. That still remains true, although our understanding of the way pollutants act on soils and vegetation has advanced somewhat since 1984.

2.1 DIRECT EFFECTS OF AIR POLLUTANTS ON PLANTS

Figure 2.1 shows that SO_2 and NO_2 together affect shoot:root ratios in barley, but could be taken to imply that more than 40 ppb (10^{12}) of the two pollutants—quite a large concentration—is needed to have a significant effect.

Exposure to SO_2 and mixtures of SO_2 and NO_2 has also been found to affect the ability of leaves to retain water when they are under water stress (Wright *et al.*, 1987). Figure 2.2 shows the rates of water loss from leaves of birch *Betula pendula* which were detached from the plant and allowed to dry. It will be seen that prior exposure to SO_2 and NO_2 had an appreciable effect on the ability to retain water under these artificial conditions. The most obvious interpretation of these curves is that cuticular transpiration is increased by exposure to pollution. An alternative explanation is that there is reduced stomatal closure during severe water stress. Recent (unpublished) work at Lancaster has shown that

appreciable effects of this nature can be detected at concentrations as low as 10 ppb SO_2 plus 10 ppb NO_2 in some species. Exposure to such concentrations is not uncommon in many parts of the British countryside. The consequences of these changes for plants under water stress in the field need to be assessed in future research.

Ozone is a powerful oxidant and reacts with leaf membranes and tissues, and the possible interactive effects of O_3 with acid rain and mists at high elevation on forests in Europe have led to considerable research on these mixtures. There is good evidence that the leakiness of cuticles is increased by O_3. Figures 2.3 and 2.4 show that O_3 and acid mists both increase the loss of magnesium from spruce needles, and the same studies show that this is also true for calcium and sulphate. Almost all experiments have shown that losses of cations in leaching water following ozone treatment of spruce needles (or indeed of other trees) are made good by root uptake. Skeffington & Roberts (1985a, b) have demonstrated that O_3 tends to increase the concentrations of the nutrients in the needles of spruce and pine. Bosch *et al.* (1986) have shown that spruce growing on a medium deficient in Ca and Mg, and treated with O_3 and acid mist, had reduced concentrations of these elements in the needles, but that if Ca and Mg had been added to the growing medium, then uptake made good the losses.

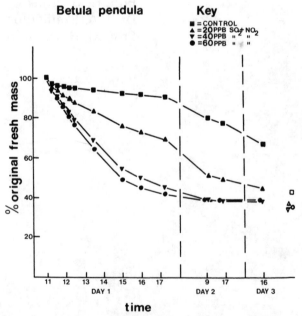

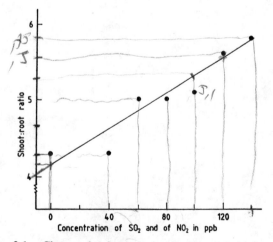

Fig. 2.1. Changes in shoot:root ratio in spring barley as a result of fumigation with SO_2 and NO_2. Seedlings were fumigated for 2 weeks, beginning 3–4 days after germination. The concentrations of both gases were equal on a unit volume basis (i.e. in ppb (10^{12})). (From Pande & Mansfield, 1985.)

Fig. 2.2. Fresh weight of excised leaves measured over time from one clone of *Betula pendula* grown for 30 days in clean air, and three rates of SO_2 plus NO_2. Each point is a mean of nine replicates, one leaf from one tree. The open symbols show the final dry mass, after oven-drying at 80°C. (From Wright *et al.* (1987).)

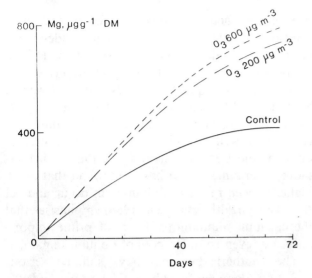

Fig. 2.3. Effect of O_3 and acid mist at pH 3·5 on the leaching of Mg from spruce needles (from Krause *et al.*, 1985).

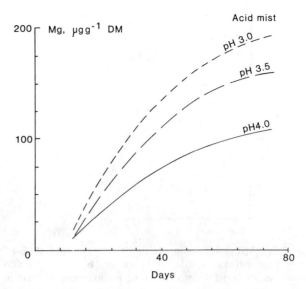

Fig. 2.4. Effect of acid mist and O_3 at $155 \, \mu g \, m^{-3}$ on the leaching of Mg from spruce needles (from Krause *et al.*, 1985).

However, the root and soil conditions in most of the experiments are rather artificial and there is also little information on the long-term effects on old trees of the low ambient concentrations of these pollutants found in remote forests.

Summer-time experiments which compare ambient and filtered air at a rural location in south-east England have demonstrated that filtration reduces leaf damage in clovers and peas which has appeared after incidents of high O_3 concentrations; furthermore, in many summers filtration has resulted in improved growth of a range of sensitive crop species (Ashmore, 1984). Recently an over-winter experiment has been carried out with red clover: after a period of very low temperatures a considerable amount of injury appeared on the foliage, but this was significantly reduced by filtration (Ashmore *et al.*, 1987). Ashmore & Dalpra (1985) have also demonstrated that a decrease in yield of peas occurs along a transect from the countryside into central London; this was tentatively ascribed to an interaction between O_3 and increasing SO_2/NO_2 concentrations into the city. However, recent filtration experiments in Glasgow and Leatherhead have also produced somewhat conflicting results, with both positive and negative effects. Such effects are probably the result of complex interactions between the pollutants present and also with various climatic factors, such as low temperatures, which are known to modify the response of plants to air pollution. The use of open-top chambers to investigate phenomena of this kind is increasing and there is better understanding of how to use them effectively—they should help to produce considerable advances in knowledge in the next few years.

There is evidence to show that some plants are more sensitive to pollutants in winter than at other times of the year. There is also good evidence to suggest that pollutants (perhaps by damaging plant membranes) may reduce frost resistance and winter hardiness. Resistance to winter desiccation, which is particularly important in evergreens, may also be impaired by pollutants due to modifications in the functions of the leaf cuticle which provides the main barrier to water loss, and interactions between pollutants and winter bleaching may occur (Davison & Barnes, 1986).

2.2 EFFECTS ON SOILS

The Watt Committee report (1984) considered that most soils could compensate for the increased loss of base cations due to acid precipitation, and that soil acidification due to the same cause was unlikely to be significant in relation to other acidifying factors. Recently, Hallbaecken & Tamm (1986) have produced evidence that the pHs of forest soils in south-west Sweden under both beech and spruce were 0·3–0·5 units lower in 1982/83 compared with 1927, with differences in the litter layer being as much as a whole unit (Fig. 2.5). They also show that under spruce, when pH is plotted against forest age, the acidifying effect of forest growth is clearly apparent, but the pH of the later samples is displaced downwards (Fig. 2.6). They interpret this difference as partly due to acid precipitation. The evidence, however, is not unequivocal, because soil

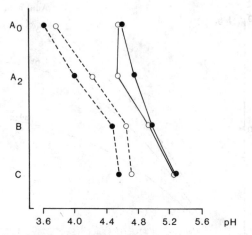

Fig. 2.5. Arithmetic mean pH in 1927 (——) and 1982–83 (----) for different soil layers in spruce (●) and beech (○) stands in south-west Sweden (from Hallbaecken & Tamm, 1986).

pH is a quantity which can vary from day to day depending largely on the amount of soil water and its salt content, and it also assumes that forestry causes no permanent acidification. There is, therefore, some concern that acid precipitation is causing significant and continuing soil acidification in Scandinavia over and above that caused by land use practices and natural processes, but it has not yet been unequivocally demonstrated.

Ulrich (1986) still maintains that soil changes are

one of the factors involved in forest decline in central and western Europe. He cites evidence for reductions of calcium and magnesium in soils from many soil types, except over calcareous rocks. However, experiments in Norway, Sweden, Scotland and Canada have indicated that the roots of conifers are able to withstand aluminium concentrations well in excess of those encountered in even the most acidic soils (Anon., 1988). Furthermore, in organic soils it has been found that much higher concentrations of aluminium occur and yet trees grow healthily. Ulrich (1986) emphasises that the calcium–aluminium ratio is of prime importance, but even so differences of opinion remain.

The southern Pennines have suffered serious pollution damage for about 200 years and the vegetation has been drastically changed over large tracts of country. Afforestation experiments in the 1950s and early 1960s failed completely, or the trees grew very badly. In many of these trials the trees which survived have now resumed growth and new experiments have been more successful (Lines, 1984). It can be inferred that the improvement follows a marked reduction in the concentrations of SO_2 and particulates over this period of time. These soils, which range from peats through surface water gleys to better drained podzols, are apparently resilient enough to resist or to recover from such

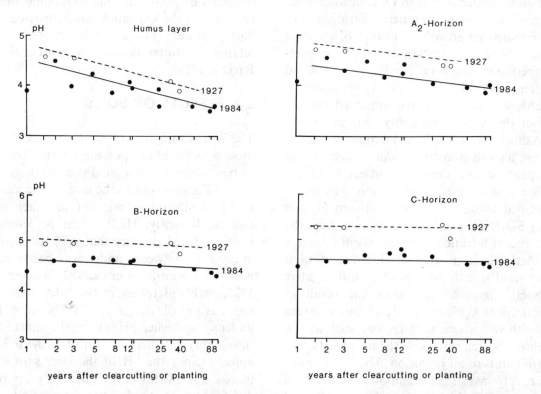

Fig. 2.6. pH of field-moist soil plotted against the logarithmic age for different spruce stands in south-west Sweden. (Simplified from Hallbaecken & Tamm, 1986.)

heavy depositions and can still support reasonable tree growth.

The increase in nitrogen pollutants in the southern Pennines in recent years and their importance in determining the growth of *Sphagnum* species has been demonstrated by Press *et al.* (1986). *Sphagnum* has also been shown to be effective in trapping mist droplets (Woodin & Lee, 1987), which frequently contain high concentrations of solutes. Polluted occult deposition may explain the decline of mountain summit bryophytes in other parts of the United Kingdom.

The limited amount of work which has been performed to establish the effects of acidifying pollutants on mycorrhizal development has suggested that roots exposed to acid treatments are associated with different mycorrhizal fungal species than roots in unexposed plants; however, we know little of the consequent effects on tree nutrition (Dighton *et al.*, 1986). The mechanisms causing such species changes are not yet known, but alterations in carbohydrate supply or in the amount of toxic aluminium in the soil may be important.

2.3 EFFECTS OF TREE CANOPIES AND OTHER PLANTS

Studies on throughfall tend to show that young conifers neutralise incoming acid precipitation whereas older conifers tend to acidify it still further. It has recently been suggested (Skeffington, 1987) that this is due to the increasing ratio of bare twigs and branches to foliage on older trees. Both foliage and bare branches collect dry deposition which is likely to be acid, but neutralisation takes place largely on foliage; and hardwoods, when in leaf, neutralise rainfall acidity.

Little is known about the effects of low canopy vegetation such as heather and bracken. In view of the extensive cover in the UK by these types of vegetation, further investigation is needed in this area. The effects of changes in soil water acidity caused by vegetation cover may be mainly local, resulting in shifts in internal cycling, or may have more serious implications for water quality of catchments.

2.4 NEUTRALISATION BY SOILS

The capacity of soils to neutralise acid precipitation is almost infinite, but the amount of neutralisation depends on the contact achieved between percolating water and the weathering rock where the neu-

tralisation is largely achieved. The properties of the upper soil horizons often determine the composition of stream water, especially during storm events, so it follows that anything which influences the upper soil horizons will affect stream water quality. These influences include land use practices as well as acid deposition.

Deposition of SO_2 is beginning to decline now as emission reductions take effect, and it has recently become clear that the speed of response of stream water will depend to a large extent on the form and content of stored soil S. Some soils in the UK (podzols of north-west Wales) have already become sulphate-saturated, leading to increases in acidity and in concentrations of aluminium in drainage waters. The importance of the sulphate input from acid deposition superimposed on the maritime-derived sulphate inputs must be established, as this additional input could significantly alter the chemistry of terrestrial and fresh water ecosystems.

Recent research has shown that the influence of wet deposition on decomposition processes in soils is likely to be important only as a result of long-term changes in soil pH. Work at the Institute of Terrestrial Ecology on the impact of simulated 'acid rain' on decomposition of forest litter in the UK has demonstrated that acidic coniferous litters have a flora and fauna well-adapted to acidic conditions, and are little affected by acid rain treatments. De-

Fig. 2.7. Spatial distribution of ammonium wet deposition in kg N ha^{-1} year^{-1} (from Buijsman & Erisman, 1986).

ciduous litters tend to be well-buffered and, again, are little affected (Ineson, 1983) unless the base reserve is limited.

2.5 DEPOSITIONS OF AMMONIA

A recent development is the realisation that concentrations of NH_3 and NH_4^+ particulates can be very high. Most work has been done in The Netherlands and this has shown that daily means of $250 \, \mu g \, m^{-3}$ can occur in cattle-rearing areas. Figure 2.7 shows that, for ammonium, wet deposition alone can reach $14 \, kg \, N \, ha^{-1} \, year^{-1}$ in the hotspots. Earlier estimates of nitrogen depositions in the UK appeared to be low because these did not take ammonium into account and new models allocate higher concentrations to animal-rearing areas. There is evidence of damage to pine trees in The Netherlands which is believed to be due to nutrient imbalance as well as fungal damage (Roelofs *et al.*, 1985).

2.6 FOREST DAMAGE IN EUROPE

Table 2.1 shows forest damage assessments for five European countries, expressed in damage classes. This table shows that the year in which damage became serious (i.e. a big increase in Class 2 damage) varies between countries; that the damage in West Germany has stabilised in the last two years with some recovery in Scots pine; and that silver fir, the species most badly affected in West Germany, with damage starting in the 1970s, has only been seriously affected in Switzerland in 1986.

Figure 2.8 shows the course of decline in silver fir in east Bavaria since 1976. A dramatic reduction occurred from 1982 onwards, and this was the year in which serious concern was first expressed about the health of Norway spruce. Thus, although the declines of silver fir and Norway spruce have been separated (Anon., 1986, and see below), circumstantial evidence strongly suggests a linking factor.

Kandler (1987) has pointed out that there is good

Table 2.1 Forest damage assessment in five European countries 1983–86 (from Innes, 1987)

	Needle/leaf loss by % and class															
	0–10% (0)				11–25% (1)				26–60% (2)				61–100% (3 + 4)			
	83	84	85	86	83	84	85	86	83	84	85	86	83	84	85	86
United Kingdom																
Sitka spruce		65	83	45		28	12	39		6	5	15		1	0	1
Norway spruce		71	84	32		26	15	36		3	1	31		1	0	1
Scots pine		49	74	25		29	18	41		16	7	32		5	1	3
West Germany																
Norway spruce	59	49	48	46	30	31	28	32	10	19	21	20	1	2	3	2
Scots pine	56	41	43	46	32	38	41	40	10	20	15	13	1	1	2	1
Silver fir	25	13	13	18	27	29	21	22	41	45	50	49	8	13	16	11
Beech	74	50	46	40	22	39	40	41	4	11	13	18	0	1	1	1
Oak	85	57	45	39	13	35	39	41	2	9	16	19	0	0	1	1
Other trees	83	69	69	65	9	24	23	25	8	7	7	9	0	1	1	1
Switzerland																
Norway spruce		65	63	50		28	29	36		6	6	12		1	2	2
Scots pine		50	35	34		31	47	43		16	13	19		1	5	4
Silver fir		62	60	47		27	28	36		9	8	13		2	4	4
Larch		64	66	39		28	23	44		7	7	12		1	4	5
Beech		74	69	52		23	27	40		3	3	7		0	1	1
Oak		71	60	37		28	33	50		1	6	11		0	1	2
Maple		86	86	73		11	11	25		2	1	1		1	2	1
Ash		84	77	57		13	20	36		3	2	7		0	1	7
Netherlands																
Norway spruce		62	48	49		28	41	34		7	9	12		3	2	4
Scots pine		34	48	50		51	36	33		12	14	13		2	2	3
Corsican pine		57	40	19		34	42	29		8	15	40		1	3	12
Douglas fir		50	33	17		39	43	27		9	22	45		2	2	11
Beech		71	72	68		24	21	26		4	6	5		1	1	2
Oak		57	40	30		38	39	42		5	19	20		1	2	9
Luxembourg																
Norway spruce		79	84	87		17	12	10		3	3	2		2	1	1
Oak		59	77	81		34	20	16		6	3	2		2	1	0
Beech		66	70	67		28	28	27		5	5	6		1	1	1

Note: The table shows, for example, that in the UK, 28% of the Sitka spruce surveyed in 1984 had lost 11–25% of their needles (i.e. were in the West German class 1).

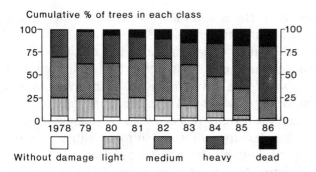

Fig. 2.8. The course of the decline of silver fir in east Bavaria in permanent observation plots (from Hoerteis & Schmidt, 1986).

documentary evidence for sites where spruce stands were in poor condition in the past (and which would now be classed as badly damaged) but which now carry healthy mature stands of spruce. He also shows evidence for recovery of individual spruce trees in recent years.

The evidence on wood increment is conflicting, but Dong & Kramer (1986) show increment reduction associated with needle loss in stands of silver fir, Scots pine and Norway spruce in Lower Saxony (Fig. 2.9). In contrast larch, with a needle loss of 18 per cent, showed no increment reduction.

The three surveys done in the United Kingdom (Binns *et al.*, 1985, 1986; Innes *et al.*, 1986), from which the UK figures in Table 2.1 have been distilled, show moderate amounts of damage in the first two years but an increase in 1986, which is believed to be largely due to winter conditions and, in Scots pine, to needle-cast fungi. Those responsible for the assessment of damage in UK forests consider that the damage observed is not unusual, though the incidence of winter damage or fungal attack may be severe. There are problems with consistency of observation and changes in perception of damage following training of observers, which, because of the subjective nature of the assessments, will occur wherever surveys of forest health are carried out. It is difficult to make valid comparisons between the UK and other countries because of the differences in growing conditions, management practices and, in some instances, tree species. Broadleaves present a particular problem and considerable work has been done in Germany, and more recently in the UK, to assess ways of looking at broadleaves in relation to their health (Lonsdale, 1986*a*, *b*). Others have considered that the health of UK trees is not satisfactory and that this is due to air pollution (Rose & Neville, 1985).

The difficulty is that most of the symptoms of

stress exhibited by broadleaved trees, and indeed by conifers, are non-specific, i.e. if a tree is stressed by drought it shows much the same symptoms as if its roots were attacked by fungi or if it had a mild dose of pollutants; furthermore, drought and high ozone concentrations are linked together. It thus becomes difficult either to eliminate pollution as a damaging agent on trees or to say that trees would not have been in better health were there no pollution at all.

Thus surveys of tree health have limitations: it is difficult to exclude observer bias; they need to be carefully designed and continued for long periods to detect change; and the symptoms are usually not characteristic of a particular stress. This has led to the investigation of a number of diagnostic tests specific to pollutant stress. These include a number of leaf surface, biochemical and physiological measurements. A recent multi-national pilot survey in Europe has shown that some of these tests can identify significant differences between trees growing in different pollution climates. This is an area requiring further research which it is hoped will lead to the incorporation of such tests into future surveys.

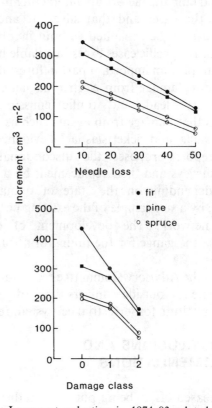

Fig. 2.9. Increment reduction in 1974–83 related to crown needle loss %, in mixed stands of silver fir, Scots pine and Norway spruce, and in a pure stand of Norway spruce, in the coastal region of Lower Saxony (from Innes, 1987, redrawn from data in Dong & Kramer, 1986).

2.7 TYPES OF DAMAGE AND CURRENT THEORIES ON THE CAUSES

The second report of the Advisory Council on Forest Damage/Air pollution of the German Government and the Länder (Anon., 1986) discusses forms of damage and possible causes. It follows Rehfuess (1985) in distinguishing several forms of damage to Norway spruce as follows:

(i) needle yellowing at high altitudes in the German highlands (Mittelgebirge);
(ii) crown thinning at middle altitudes in the highlands;
(iii) needle reddening of older stands in south Germany;
(iv) yellowing at high altitudes in the calcareous Alps;
(v) crown thinning in north coastal areas.

The report regards silver fir decline in south Germany and beech decline in north Germany as different phenomena. It concludes that pollution is an important factor in four of the five types of spruce decline, although its role in needle reddening appears less certain. The report considers that ozone and climatic factors are interacting in at least some of the cases and that silver fir and beech damage cannot be explained without involving air pollution. In specific cases excess available nitrogen appears to play an important part in forest damage. It rules out damage from electro-magnetic waves, radioactivity, lead, and traffic emissions, and considers that damage from agents such as viruses, mycoplasmas, and rickettsias lacks evidence.

The report proposes the discontinuation of annual surveys and their replacement by a flexible system depending on the state of damage, for example, by a sub-sample of the existing permanent sample network. The development of damage should be the gauge for the timing of a full inventory.

Finally the Advisory Committee recommends the establishment, outside universities, of a West German institute for terrestrial ecosystem research.

2.8 CONCLUSIONS AND RECOMMENDATIONS

The increased effort being put into studies of the effects of pollutant mixtures is beginning to produce a better understanding of how they act, but there is still a long way to go, in particular in elucidating the long-term effects of modest concentrations of pollutant mixtures and their interactions with climatic stress, especially frosts and droughts.

It is widely accepted that forest decline is not a single syndrome and there is less synchrony (for example, with silver fir) in its incidence than was previously thought. There are also some doubts about the novelty of some kinds of damage. Nevertheless, the appearance of so many ailments in the forest at roughly the same time suggests a linking or triggering factor. As more information on the pollutants present in the areas damaged becomes available, this should also help in improving understanding.

The increasing use of open-top chambers to investigate the effects of pollutants should help to improve our understanding of the relationships and this work should be encouraged, particularly in collaboration with physiological studies. Furthermore the interactions between cold stress and pollutant damage, and their consequent effects on leaf function, need further work.

The search for diagnostic tests to detect pollutant stress has only just begun and, though the way in which such tests might be used in practice is not yet clear, the work should in due course considerably increase our understanding of stressing mechanisms.

Large-scale surveys of forest damage have only limited usefulness (one of which may be political) and there is a shift of opinion, in Germany at least, away from them and towards more detailed work on a sub-sample of long-term plots.

REFERENCES

Anon. (1986). *Forschungsbeirat Waldschäden/Luftverunreinigung der Bundesregierung und der Länder. 2 Bericht.* Kernforschungszentrum, Karlsruhe.

Anon. (1988). *The Effects of Acid Deposition on the Terrestrial Environment in the UK*, Terrestrial Effects Review Group Report. HMSO, London.

ASHMORE, M. R. (1984). Effects of ozone on vegetation in the United Kingdom. In *The Evaluation and Assessment of the Effects of Photochemical Oxidants on Human Health, Agricultural Crops, Forestry, Materials, and Visibility. Proc. Int. Workshop March 1984*, ed. P. Grenfeldt. Swedish Environmental Research Institute, Goteborg, pp. 92–104.

ASHMORE, M. R. & DALPRA, C. (1985). Effects of London's air on plant growth. *London Environmental Bulletin*, **3**, 4–5.

ASHMORE, M. R., MIMMACK, A., MEPSTED, R. & BELL, J. N. B. (1987). Research at Imperial College with open-top chambers, 1976–1986. In *Microclimate and Plant Growth in Open-Top Chambers*, Commission of the European Communities, Air Pollution Report 5, pp. 98–101.

BINNS, W. O., REDFERN, D. B., RENNOLLS, K. & BETTS, A. J. A. (1985). Forest health and air pollution: 1984 survey. *Forestry Commission Research and Development Paper*, No. 142.

BINNS, W. O., REDFERN, D. B., BOSWELL, R. & BETTS, A. J. A. (1986). Forest health and air pollution: 1986 survey. *Forestry Commission Research and Development Paper*, No. 147.

BOSCH, C., PFANNKUCH, E., REHFUESS, K. E., RUNKEL, K. H., SCHRAMEL, P. & SENSER, M. (1986). Einfluss einer Duengung mit Magnesium und Calcium, von Ozon und saurem Nebel auf Frosthaerte, Ernaehrungszustand und Biomasseproduktion junger Fichten (*Picea abies* (L.) Karst.). *Forstwissenschaftliches Centralblatt*, **105**, 218–29.

BUIJSMAN, E. & ERISMAN, J. W. (1986). *Ammonium flux in Europe*. Instituut voor Meteorologie en Oceanografie, Rijksuniversiteit, Utrecht. R.86.5.

DAVISON, A. W. & BARNES, J. D. (1986). *Effects of winter stress on pollutant responses. CEC COST Workshop*. Roskilde, Denmark.

DIGHTON, J., SKEFFINGTON, R. A. & BROWN, K. A. (1986). The effects of sulphuric acid (pH 3) on roots and mycorrhizas of *Pinus sylvestris*. In *Mycorrhizae; Physiology and Genetics*, ed. V. Gianinazzi-Pearson & S. Gianinazzi. INRA, Paris, pp. 739–43.

DONG, P. H. & KRAMER, H. (1986). Auswirkungen von Umweltbelastungen auf das Wuchsverhalten verschiedener Nadelbaumarten im nordwestdeutschen Kuestenraum. *Der Forst- und Holzwirt*, **41**, 286–90.

HALLBAECKEN, L. & TAMM, C. O. (1986). Changes in soil acidity from 1927 to 1982 in a forest area of south-west Sweden. *Scandinavian Journal of Forest Research*, **1**, 219–32.

HOERTEIS, J. & SCHMIDT, A. (1986). Die Entwicklung des Gesundheitszustandes der Weisstanne auf 10 Beobachtungsflächen in Ostbayern. *Der Forst- und Holzwirt*, **41**, 580–2.

INESON, P. (1983). The effect of airborne sulphur pollutants upon decomposition and nutrient release in forest soils. PhD thesis, University of Liverpool.

INNES, J. (1987). Air pollution and forestry. *Forestry Commission Bulletin*, No. 70.

INNES, J. L., BOSWELL, R., BINNS, W. O. & REDFERN, D. B. (1986). Forest health and air pollution: 1986 survey. *Forestry Commission Research and Development Paper*, No. 150.

KANDLER, O. (1987). Klima und Baumkrankheiten. In *Klima und Witterung im Zusammenhang mit den neuartigen Waldschäden. Proc. Symposium 13/14 Oct 1986*, GSF Muenchen, Bericht 10, pp. 269–75.

KRAUSE, G. H. M., JUNG, K. D. & PRINZ, B. (1985). Experimentelle Untersuchungen zur Aufklaerung der neuartigen Waldschäden in der Bundesrepublik Deutschland. In *Waldschäden. Einflussfaktoren und ihre Bewertung, Kollo-*

quium Goslar, 18–20 June 1985, ed. H. Stratmann. *VDI-Berichte*, **560**, 627–56.

LINES, R. (1984). Species and seed origin trials in the industrial Pennines. *Quarterly Journal of Forestry*, **78**, pp. 9–23.

LONSDALE, D. (1986a). Beech health study 1985. *Forestry Commission Research and Development Paper*, No. 146.

LONSDALE, D. (1986b). Beech health study 1986. *Forestry Commission Research and Development Paper*, No. 149.

PANDE, P. C. & MANSFIELD, T. A. (1985). Response of spring barley to SO_2 and NO_2 pollution. *Environmental Pollution, Ser. A*, **38**, 87–97.

PRESS, M. C., WOODIN, S. J. & LEE, J. A. (1986). The potential importance of an increased atmospheric nitrogen supply to the growth of ombrotrophic *Sphagnum* species. *New Phytologist*, **103**, 45–55.

REHFUESS, K. E. (1985). On the causes of Norway spruce (*Picea abies* Karst.) decline in Central Europe. *Soil Use and Management*, **1**, 30–2.

ROELOFS, J. G. M., KEMPERS, A. J., HOUDIJK, A. L. F. M. & JANSEN, J. (1985). The effect of air-borne ammonium on *Pinus nigra* var. *maritima* in The Netherlands. *Plant & Soil*, **84**, 45–56.

ROSE, C. & NEVILLE, M. (1985). *Final Report: tree dieback survey*. Friends of the Earth, London.

SKEFFINGTON, R. A. (1987). Do all forests act as sinks for air pollutants? Factors influencing the acidity of throughfall. In *Acidification and water pathways. UNESCO/IHP Symposium, Bolkesjo, Norway*, Vol. II, pp. 85–94.

SKEFFINGTON, R. A. & ROBERTS, T. M. (1985a). The effects of ozone and acid mist on Scots pine saplings. *Oecologia*, **65**, 201–6.

SKEFFINGTON, R. A. & ROBERTS, T. M. (1985b). Effects of ozone and acid mist on Scots pine and Norway spruce —an experimental study. *VDI-Berichte, Düsseldorf*, **560**, 747–60.

ULRICH, B. (1986). Die Rolle der Bodenversauerung beim Waldsterben: langfristige Konsequenzen und forstliche Moeglichkeiten. *Forstwissenschaftliches Centralblatt*, **105**, 421–35.

Watt Committee (1984). *Report No. 14—Acid Rain*. The Watt Committee on Energy, London.

WOODIN, S. J. & LEE, J. A. (1987). The fate of some components of acid deposition in ombrotrophic mires. *Environmental Pollution*, **45**, 61–72.

WRIGHT, E. A., LUCAS, P. W., COTTAM, D. A. & MANSFIELD, T. A. (1987). Physiological responses of plants to SO_2, NO_2 and O_3: implications for drought resistance. In *Direct Effects of Wet and Dry Deposition on Forest Ecosystems. CEC COST 612 Workshop, Loekeberg, Sweden, October 1986*, pp. 187–200.

Section 3

Freshwater

Desmond Hammerton

Director, Clyde River Purification Board, Glasgow

This paper presents the work of a sub-group of the Watt Committee working group on Air Pollution, Acid Rain and the Environment.

Membership of Sub-group

D. Hammerton (Chairman)

Dr R. W. Battarbee
T. Carrick
D. H. Crawshaw
Dr A. P. Donald
R. Harriman
A. V. Holden
Prof. K. Mellanby
Dr P. S. Maitland
Dr S. Warren

3.1 INTRODUCTION

This section of the report documents the progress which has been made since our first report (Watt Committee, 1984) in assessing the impact of acid deposition on the freshwater environment. Considerable advances have been made during the past four years in understanding the history and causes of acidification, particularly in its more acute phase between 1930 and 1970. Firmer conclusions have now been drawn concerning the extent of acidification in Britain, the part played by anthropogenic emissions, the effect of afforestation in vulnerable areas and the reversibility of acidification processes.

The two reports should be read together, the former serving as a convenient starting point, particularly with regard to acidification mechanisms and techniques for assessing acidification which have not been described in detail in this report. The report has, of necessity, condensed much of the source material but the detailed references provide a useful guide to the published research.

3.2 RECENT RESEARCH

3.2.1 Sensitivity to acidification

3.2.1.1 Geology
Only in certain areas of the country are surface waters acidified or sensitive to acidification. Such areas are those where geological weathering rates are slow and where soils are already acid and have little ability to neutralise any further increase in acidity.

The precise distribution of acidified or sensitive surface waters in the UK is not yet known, but a useful first approximation can be gained from a map of groundwaters prepared by the British Geological Survey (cf. Edmunds & Kinniburgh, 1986) in which four sensitivity categories are defined. These areas coincide with the granites and acid igneous and metamorphic rocks of Highland Scotland, the granites of Galloway, Cumbria and South-west England, the gritstones of the English Pennines, and the lower Palaeozoic sedimentary and metamorphic rocks of Wales. The Cretaceous sandstones that outcrop in Central and South-east England are in the medium category, but most rocks and soils in England and in Central and Eastern Scotland are capable of neutralising acid deposition very effectively, and are regarded as non-sensitive (see Fig. 3.1).

3.2.1.2 Catchment characteristics
Although the basic geological characteristics of a region control the general sensitivity of that region to acid deposition, the acidity of any one stream or lake is also influenced by the individual features of the stream or lake catchment. Variations in microgeology, in the presence and type of superficial deposits, in the ratio of organic to mineral soils, in the pattern of hydrological flows, and in the use made of the catchment by man, can all affect the response or lack of response of a water body to acid deposition. Such variations between catchments explain why lakes on apparently identical geology in the same region can respond differently to acid deposition. For example, afforestation enhances acidification (Harriman & Morrison, 1982; Stoner *et al.*, 1984) whilst agricultural liming may prevent or delay acidification.

3.2.1.3 Acid deposition
Acidifying substances, mainly sulphates, nitrates and hydrogen ions, are removed from the atmosphere by wet and dry deposition processes (see Section 1). Contours for the mean annual pH of precipitation in the UK are shown on Fig. 3.1. The most acid areas coincide with the most densely populated and industrialised parts of the country. But the total amount of acidity deposited increases to the West and North-west because of the high precipitation in these areas. A similar pattern occurs for sulphate and nitrate deposition.

3.2.1.4 Affected areas
It is not yet certain what the critical levels of deposition are although some authors have suggested that approximately $15 \, \text{kg}$ wet $SO_4 \, \text{ha}^{-1} \, \text{year}^{-1}$, equivalent to an average pH of precipitation of about 4·6–4·7, is the limit for damage to sensitive aquatic ecosystems. On this basis, with the exception of the north-west of Scotland, almost all geologically sensitive areas shown in Fig. 3.1 are likely to have been acidified to a greater or lesser extent.

3.2.2 Surface water acidification: trends and spatial patterns

3.2.2.1 Sediment records
Trends in lake acidification can be inferred with a considerable degree of confidence from the changes in the composition of fossil diatom assemblages preserved in lake sediments, and a simple model that relates modern diatom assemblages to

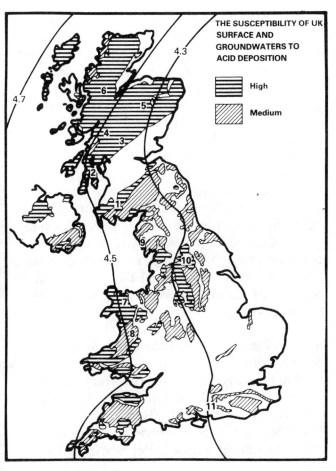

Fig. 3.1. Susceptibility of UK surface and groundwaters to acid deposition (from Edmunds & Kinniburgh, 1986), contours of annual mean pH of precipitation (from Barrett *et al.*, 1983), and key research areas.

Explanation. Shaded areas are those where surface waters are acidified or are most susceptible to acidification as defined by Edmunds & Kinniburgh (1986). Unshaded areas are those where surface waters have little or no susceptibility to acidification. The classification is mainly based on bedrock geology. pH contours are extrapolated from Barrett *et al.*, (1983) using data from EMEP and from Donald & Stoner (in press). Values for the annual mean pH of precipitation vary from year to year. In 1986, for example, values in Scotland were up to 0·2 pH units higher than shown here. The main acidified areas are those on sensitive geology within the pH 4·5 contour but local variations in geology, soils, land-use and altitude limit the accuracy of this generalisation.

Key areas are:

SCOTLAND

1. *Galloway.* Diatom analysis shows that many acidified lakes occur in the Galloway region. Those now fishless include L. Enoch, L. Valley, L. Neldricken, L. Arron, and L. Fleet. Lakes with substantially reduced fish populations include L. Riecawr, L. Grannoch, L. Dee and L. Trool. Dipper populations are sparse. Aquatic macrophyte resurveys show decline and disappearance of some *Potamogeton* species and invasion of *Sphagnum* in the submerged zone of acidified lakes. Affected sites occur on, or partly on, granitic rocks in the upland region and problems are severely exacerbated by afforestation.
References: Flower & Battarbee (1983); Battarbee (1984); Burns *et al.* (1984); Battarbee *et al.* (1985); Raven (1985, 1986); Howells & Brown (1986); Flower *et al.* (1987); Harriman *et al.* (1987); Maitland *et al.* (1987); Vickery pers. comm.

2. *Arran.* Diatom analysis of a core for Loch Tanna shows acidification by approximately 0·5 pH units, suggesting that all surface waters on the granitic part of the island are similarly affected.

3. *Trossachs.* Streams and lakes with afforested catchments in the Loch Ard area are acidified and have poor fisheries. Loch Tinker, a lake with a moorland catchment, has been slightly acidified, suggesting that surface waters on the meta-sediments in this area are severely affected only when catchments are afforested.
References: Harriman & Morrison (1982); Harriman & Wells (1985); Kreiser *et al.* (unpublished).

4. *Rannoch Moor.* Loch Laidon has been slightly acidified but no fish decline has been reported. Part of Loch Laidon lies within a National Nature Reserve.
References: Harriman & Wells (1985); Flower *et al.* (in prep).

5. *Cairngorms.* There are some naturally fishless lakes but fish declines have been reported in the Lochnagar area. So far diatom analyses in progress suggest acidification of the corrie lochs.
References: Freshwater Fisheries Laboratory (1982–1984); Flower *et al.* (in prep.); Jones *et al.* (in prep.).

6. *North-west Scotland.* Most lochs and streams are relatively unacidified and fish populations are good. Diatom data are not yet available, but except for some very acid brown water lochs, such as Long Loch on Dunnet Head, few waters with pH < 5·5 occur.
Reference: Harriman & Wells (1985); Maitland *et al.* (1987).

WALES

7. *North Wales.* A large number of acidified lakes occur in the Migneint and Harlech dome areas and include L. y Bi, L. cwm Mynach, L. Dulyn, L. Llagi, L. Conwy, and L. Gamallt. L. Conwy and L. Gamallt have lost their trout populations. Dippers along streams with pH < 6·0 are significantly fewer and have a poorer breeding performance than along less acid waters.
References: Ormerod *et al.* (1986); Patrick & Stevenson (1986); Stevenson *et al.* (1987); Ormerod & Tyler (1987).

8. *Mid-Wales.* Acidified and fishless lakes occur in the Upper Ystwyth, Upper Teifi and Upper Tywi catchments. They include L. Pendam, L. Blaenmelindwr, L. Hir, L. Berwyn, and L. Brianne. Dipper populations influenced as for North Wales.
References: Stoner *et al.* (1984); Stoner & Gee (1985); Fritz *et al.* (1986); Kreiser *et al.* (1986); Ormerod *et al.* (1986); Ormerod & Tyler (1987); Stevenson *et al.* (1987); Underwood *et al.* (1987).

ENGLAND

9. *Cumbria.* Acidification problems are mainly associated with the Eskdale granites. There is diatom evidence for the acidification of Scoat Tarn, and Levers Water is now fishless. The large lakes appear to be unaffected.
References: Sutcliffe *et al.* (1982); Sutcliffe (1983); North-West Water (1986); Sutcliffe & Carrick (1986); D. H. Crawshaw (pers. comm.); T. R. Carrick (pers. comm.); E. Y. Haworth (pers. comm.).

10. *Pennines.* Some of the most acid waters in Britain occur in the Pennines, but the extent of acidification is hindered by the lack of appropriate historical and sedimentological records.
References: North-West Water (1986).

11. *South-east England.* Acid and acidified sites mainly coincide with areas of Tertiary sandstone. Former natterjack toad breeding sites that are thought to have been acidified include Pudmore Pond, Cranmer Pond, and Bordon West. Major declines have occurred at Woolmer Pond, where breeding only occurs in ponds with artificial Ca enrichment from concrete.
References: Beebee (1976, 1977); Beebee & Griffin (1977).

measured pH values can be used to reconstruct past pH values. An example of this approach using data from the Round Loch of Glenhead, together with a reconstructed pH profile is shown in Fig. 3.2. The sediment is dated using the ^{210}Pb method.

So far eight lakes in Galloway, two on Rannoch Moor, one in Cumbria, and four in Wales have been studied using this method. The results indicate that lakes on sensitive geologies in all these areas have experienced a significant decrease in pH in recent (post-1800) times. So far analyses have not been carried out for sites in the north-west of Scotland where acid deposition is quite low. However, analyses have been carried out in such areas of sensitive geology but with low acid deposition in Norway, Finland, Austria and the USA, and in none of these cases has recent acidification been observed (Batterbee & Charles, 1986).

3.2.2.2 Trends in water chemistry: UK regional data sets
The United Kingdom Acid Waters Review Group has obtained from Regional Water Authorities, Scottish River Purification Boards and other organisations data relating to waters considered sensitive to acidification.

Very few records are available of regular measurements made over a period of decades and those data which were obtained were far from ideal for use in the assessment of chemical trends. The records generally related to the past 10–15 years, a period when sulphur emissions were actually declining. Not all of the 100 data sets examined included determinands other than pH and most of the measurements had been made in the lower reaches of rivers, where low pH values are unlikely. Furthermore, pH measurements may not always have been made with the precautions necessary in waters of low ionic strength for the purposes of this study.

The results show that downward trends in pH are not widespread in the areas for which data were obtained. On the other hand the results must be regarded as inconclusive because of the limitations listed above.

3.2.2.3 Trends in water chemistry: international comparisons
During the early eighties the concentration of free acidity (H^+) in bulk precipitation in Scotland has declined by about 50% (Harriman and Wells, 1985). Sulphate concentrations have also declined but by a smaller amount. Reports from Sweden

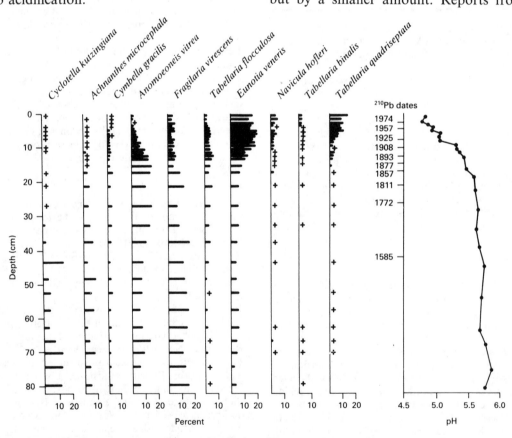

Fig. 3.2. Diatom diagram and pH reconstruction for the Round Loch of Glenhead, Galloway, South Wear, Scotland (from Flower & Battarbee, 1983).

indicate a 30% reduction in sulphate for areas of central Sweden. In Canada most attention has been paid to the effect of reduced SO_2 emissions from the Sudbury smelter.

In the UK there is now evidence of reversibility of acidification in the Galloway lakes (Battarbee *et al.*, 1988). In Sweden, Forsberg *et al.* reported a 20–50% decline in sulphate concentrations for a number of lakes and a maximum 47% decline in H^+ concentration in two lakes. Lazerte & Dillon reported a 30% decline in sulphate levels in Clearwater Lake between 1978 and 1983 (545 to 379 μeq litre^{-1}) and a 57% decline in H^+ (4·23 to 4·60 μeq litre^{-1}).

In a comparison of Ontario lakes between the period 1974–76 and 1981–83, Keller & Pitblado reported a general decline in acidity and sulphate between the two periods with the degree of change related to distance from the major SO_2 emission at Sudbury.

Although data are generally fragmented and cover a variable timescale the indication is that reductions in emissions cause improvements in water chemistry in a relatively short period of time.

3.2.3 Causes of acidification

3.2.3.1 Natural acidity
Naturally acid waters (pH below 5·6) occur in the UK as small, shallow pools and headwater streams where the catchment areas are dominated by peatlands. The acidity is caused by organic acids derived from the peats. Such waters are usually highly coloured, and most, but not all, are fishless. In areas of high acid deposition it is probable that these systems have been further acidified.

Except for these coloured waters there are very few continually acid (below pH 5·6) surface waters in areas of low acid deposition although some natural processes can cause short-lived acid episodes in waters of low alkalinity. These include soil oxidation processes and the results of sea-salt intrusions (Harriman & Wells, 1985).

3.2.3.2 Long-term acidification
Some of the acidity in acid lakes is the result of soil and vegetation changes that have or may have taken place over the time period of the post-glacial period (Pennington, 1984). In the UK most natural lakes are between 10 000 and 13 000 years old and were formed from depressions remaining after the recession of the last ice-sheet. Diatom analysis of the earliest sediments in such lakes shows that many

upland lakes were much more alkaline during the first few thousand years of their history (Round, 1957; Pennington *et al.*, 1972), and subsequently became more acid as organic soils developed in their catchments. However, in none of the lakes so far studied did the pH fall to values below about 5·6, and in almost all cases there has been little or no further acidification of these lakes until the post-AD 1800 phase. In one lake in Galloway, the Round Loch of Glenhead, the pH of the lake was between 5·5 and 6·0 throughout its history until a rapid acidification after 1850.

Although many lakes became more sensitive to acidification because of the loss of alkalinity during these early stages of lake history there is no evidence yet that slow progressive acidification associated with natural processes is responsible for the current biological problems.

3.2.3.3 Land use and management (excluding afforestation)
It has been suggested (Rosenquist, 1978; Krug & Frink, 1983) that a decrease in burning and grazing of lake catchments can cause soil and surface water acidification as acid heathland species (especially *Calluna vulgaris*) replace herbaceous vegetation. It has also been suggested that a decline in the use of agricultural limestone in upland areas since about 1960 may also have caused acidification (Ormerod & Edwards, 1985).

These mechanisms are unlikely to be important in the United Kingdom for several reasons:

(i) Most acidified sites occur in areas where soils have been highly acid for millennia;
(ii) In many cases (cf. Battarbee *et al.*, 1985) there is no evidence that there has been a decline in burning and grazing;

Table 3.1 An example of the chemical characteristics of waters which are (a) naturally acidified, (b) acidified by acid depositions, and (c) acidified by a combination of (a) and (b)

	pH	Ca[a]	Al-L[b]	Al-NL[c]	DOC[d]
(a) Long Loch (Caithness)	4·73	65	17	50	6·0
(b) Loch Enoch (Galloway)	4·60	32	110	1	<1
(c) Kelty Water (Loch Ard)	4·40	36	61	85	6·5

[a] Excluding sea-salts (μeq litre^{-1}).
[b] Labile monomeric aluminium (μg litre^{-1}).
[c] Non-labile (organic aluminium (μg litre^{-1})).
[d] Dissolved organic carbon (mg litre^{-1}).

(iii) Acidification occurs at sites where burning and grazing has been minimal, e.g. Loch Enoch;

(iv) Studies of early and mid post-glacial lake sediments show that the development of catchment peatlands 5000 or so years ago, and obviously prior to the impact of acid deposition, did not cause the kind of acute acidification observed in the nineteenth and twentieth centuries;

(v) Acidified areas so far identified in Wales and Scotland are within the moorland core where agricultural liming has not been practised (Patrick, 1987);

(vi) Where a decline in liming has been associated with surface water acidification as in parts of Cumbria, it is unlikely that the underlying cause is acid deposition. Liming, whilst it is still practised, merely buffers the catchment from acid deposition.

3.2.3.4 Afforestation

Areas of productive forest now account for 10% of the total land area of the UK, representing a twofold increase since the formation of the Forestry Commission in 1919. Post-war planting has been almost exclusively of coniferous species.

In addition to the reduction in water yield and the increase in sediment loads associated with afforestation, a problem of surface water acidification has been identified in afforested catchments on base-poor geologies in Scotland and Wales. Harriman & Morrison (1982) and Stoner et al. (1984) have clearly shown that stream catchments with mature forest are more acid, contain higher levels of toxic dissolved aluminium, and consequently support a poorer fishery by comparison with adjacent moorland catchments. Similar conditions have led to the impoverishment or disappearance of natural salmonid fisheries from some lakes in mid Wales and south-west Scotland, and studies of lake sediments in both areas are beginning to confirm the importance of forestry as a factor in acidification (Anderson et al., 1986).

Various mechanisms have been suggested to account for the observed differences between the quality of surface waters draining moorland and afforested catchments, including both the enhancement of atmospherically deposited acidity and processes related to tree growth and forestry drainage. At present it is not possible to state which mechanism is dominant, but the most likely process is the enhanced scavenging of dry-deposited pol-

lutants by the forest. Preliminary results from the Llyn Brianne area in mid-Wales indicate that the conifer canopy enhances the acidity of precipitation under certain meteorological conditions. When westerly wind directions were dominant, throughfall under mature sitka spruce was found to have a similar pH to that of precipitation. By contrast, both throughfall and stemflow were found to be more acidic when easterlies prevailed, suggesting the influence of dry deposition (Hornung et al., unpublished). During snow-melt in early March 1986, after a prolonged cold spell characterised by easterly winds and standing snow (mean pH 3·2), continuous monitoring of a moorland stream revealed that pH fell by 1·2 units in six hours. A similar though enhanced effect was observed in a stream draining mature conifer forest, where pH fell by 2 units in ten hours (Brown & Ormerod, unpublished).

3.2.3.5 Acid deposition

(a) *Mechanisms.* Empirical models and studies of lake sediments (Battarbee, 1984; Battarbee & Charles, 1986) suggest that the acidification of non-afforested streams and lakes in the UK is the result of acid deposition.

Whilst naturally acid lakes receive their acidity from the breakdown of organic material in their catchments, the mechanism of acidification by acid deposition is thought to be mainly due to the elevated sulphate concentration of precipitation. In sensitive catchments the movement of this anion through soil systems is electrically balanced by acid rather than base cations producing surface waters which are relatively clear and have high acidity, aluminium, sulphate and trace metal levels. So far nitrate has not been important but if the input to soils exceeds uptake, nitrate levels will begin to increase in streams and lakes, especially during snow-melt. The time-scale and rate of these processes is likely to vary from catchment to catchment depending on the sulphate adsorption capacity of the soils, mineral weathering rates, soil depth, and hydrological pathways.

(b) *Medium-term responses (the lake sediment record).* Lake sediment studies can place the acidification process in a wider spatial and temporal context. The spatial distribution of acidified lakes so far studied (Galloway, Cumbria, West Wales, and Rannoch Moor) is coincident with areas of high sulphate deposition, acid soils and sensitive geology. Control sites in areas of low sulphate deposition have yet to be examined in detail but preliminary observations suggest that strongly acidified clear water lakes do not occur in such areas. Moreover, in most cases where acidification of non-afforested sites has been identified, its beginning occurs in the middle or late nineteenth century, and accelerates in the mid-twentieth century, a time-scale that is consistent with the trends in SO_2 emissions that have been reconstructed for the last 150 years by Barrett *et al.* (1983). A further indication that this process is controlled by acid deposition is the fact that non-afforested sites studied in Galloway and Rannoch Moor are not becoming more acidic, and in some cases there is evidence of a small reversal in trends. This again would be consistent with the observed downturn in UK emissions and deposition since 1970 (Harriman & Wells, 1985).

For afforested sites the sediment evidence suggests that the most sensitive areas were acidified prior to afforestation (Flower & Battarbee, 1983; Flower *et al.*, 1987), but that less sensitive catchments have been and are continuing to be acidified following canopy closure. The continuing acidification of streams and lakes with afforested catchments indicates that a much greater reduction in sulphate deposition is required to counter the forest effect.

These spatial and temporal patterns observed in the UK (cf. Fig. 3.1) can also be observed on a larger scale in Western Europe and North America. On both continents evidence for acidification outside the area of acid deposition is meagre despite the extreme sensitivity of such sites in, for example, the Boundary Areas of Northern Minnesota or the Trondelag area of Norway (Battarbee & Charles, 1986).

Lake sediment studies also indicate that alternative hypotheses for lake acidification are usually invalid (see 3.2.3.3 and 3.2.3.4 above) and show that even very isolated upland lakes are not pristine. Geochemical analyses quite clearly show strong post-Industrial Revolution contamination by trace metals, especially Pb, Zn and Cu, and by high concentrations of carbonaceous spherules that are derived from combustion of coal and oil (Battarbee *et al.*, 1985; Battarbee *et al.*, in prep.).

(c) *Short-term responses* (*episodes*). Additional problems can be created by storms and snow-melt episodes since these can cause rapid short-term changes in stream-water chemistry. Where there has been a loss of surface water alkalinity as a result of acid deposition over the last 150 years the high levels of acidity and aluminium generated by these events can have a toxic effect on fish and other biota.

The key parameters are pH, aluminium (labile) and calcium but there has been little research into the effect of episode intensity and duration. Recent work by Rosseland & Skogheim in Norway suggests that frequent episodes of relatively low levels of acidity and aluminium could be more toxic than one episode of high acidity and aluminium. During any one episode the concentration of calcium will change in relation to pH and aluminium. Sea-salt episodes can produce very low pH levels but at the same time calcium and sodium levels are high, thus making interpretations of survival experiments very difficult.

3.2.4 Impact on biota

3.2.4.1 Changes in flora

Evidence for floristic change in acidified lakes comes from a comparison of contemporary floras in acid and non-acid lakes, from resurveys of sites, from the sediment record, and from experimental manipulation of enclosures or whole lakes.

All these sources show that acidification causes major changes in the composition of planktonic and benthic algal communities. In the plankton most diatoms disappear as the pH falls below 5·5, to be replaced by cryptophytes, dinoflagellates and chrysophytes. So far, however, there is no clear indication of a decline in planktonic productivity. In most acid systems the periphytic algal communities are more important where acidification causes a shift to acidophilous and acidobiontic diatoms and to the emergence of the filamentous green alga *Mougeotia* as a dominant taxon. Changes in the structure of aquatic macrophyte communities are less clear-cut although there is abundant evidence from many countries of the expansion of aquatic *Sphagna* in acidified waters. In the UK the best data are from Loch Fleet, where *Sphagnum auriculatum* now dominates the community at a water depth of 2–4 m but was not recorded by West & West in their survey of 1905 (West, 1910; Raven, 1985, 1986). It is also likely that other macrophytes are lost. For example, Raven (1985) showed that *Potamogeton lucens* is much less common in Galloway lakes today as compared with the West & West survey (Fig. 3.1).

3.2.4.2 Changes in fauna

(a) *Invertebrates.* There are numerous data sets concerning the invertebrate communities of streams in many parts of Great Britain, some of them covering a time period of over 25 years. However, relatively few relate to waters sensitive to acidification and the majority therefore are not relevant to the present study. Certainly this appears to be true as far as long-term data and trends in time are concerned. Analyses of invertebrate communities from streams in different parts of the country have shown that there are differences between the communities of acid and alkaline waters. In general the former are dominated by certain insects, Plecoptera, Trichoptera and Diptera, and the latter by Ephemeroptera, Mollusca and Crustacea.

Harriman & Morrison (1982) showed that in a group of streams which they studied similar numbers were present but Ephemeroptera nymphs were absent or rare in streams with low pH and high dissolved aluminium levels. Stoner *et al.* (1984) have also demonstrated that a number of groups are scarce or absent in soft or acidic waters in Wales, and experimental evidence from an artificially acidified stream has shown that invertebrate drift is especially marked where elevated levels of both hydrogen ion and aluminium are found (Ormerod *et al.*, 1987). Using TWINSPAN, results showing impoverishment of invertebrate communities have been obtained for streams in several parts of the UK. Thus there are now considerable data indicating that acidified communities do differ from those in other waters.

(b) *Fish.* There is now convincing evidence that fish stocks (especially salmonids) in a number of places in the UK have been affected by acidification. Harriman & Morrison (1982) showed that in the Trossachs area (Fig. 3.1) salmonids were absent from forest streams but present in adjacent moorland streams and that these differences were related to the hydrogen and aluminium content of the streams. Similar data were obtained by Gee & Stoner in Wales. Egglishaw *et al.* have recently shown from an analysis of the Scottish salmon catch statistics that the recent decline in the numbers of salmon caught in various catchments is highly correlated with the extent of afforestation of the catchments concerned.

In Galloway (Fig. 3.1) many lochs which previously had substantial trout populations are now fishless and there is convincing evidence that this is related to recent acidification (Harriman *et al.*, 1987). In addition, a number of lochs with pH < 5·5 appear to have declining trout populations, as indicated by the absence of juveniles in nursery streams and the absence of recent year classes in lochs. In Cumbria (Fig. 3.1) there have been significant mortalities of juvenile and adult salmonids in the rivers Duddon and Esk and these have been related to short-term episodes of acid water following storms.

(c) *Amphibians.* There is considerable evidence of a decline in recent years in the numbers of several species of Amphibia in Great Britain. Different Amphibia vary in their ability to tolerate low pHs, but although some are able to survive in bog pools at around pH 4 others cannot survive such extremes. It has been suggested by Aston (see Cummins, 1986) that this may be due to the varying amounts of dissolved aluminium in the waters concerned. The greatest problem concerns the natterjack toad *Bufo calamita*, which is a rare and protected species. It has disappeared from a number of heathland sites in south-east England probably as a result of the acidification of its breeding sites (Fig. 3.1). Beebee (1986) has shown the vulnerability of spawn and small tadpoles to acid waters. Further research in this area is needed.

(d) *Birds.* There is evidence from Wales (Fig. 3.1) of a decline in the numbers of dippers *Cinclus cinclus* on streams believed to have undergone acidification (Ormerod *et al.*, 1986, Ormerod & Tyler, 1987). Similar studies are in progress in south-west Scotland to determine whether dippers and two other bird species have been affected in this sensitive area.

3.2.4.3 Response to recent decreases in acid deposition

Between 1970 and 1985 there was a decline in UK SO_2 emissions and a corresponding decline in SO_4 deposition (Harriman & Wells, 1985). The response of surface water chemistry and freshwater biota to these changes is of considerable interest. Although there is some evidence from the diatom record that lakes with non-afforested catchments are not becoming more acidic, and in some cases may be improving, it is probably too early to assess the response of fish stock and invertebrates. Recolonisation depends not only on the establishment of an improved chemical environment but on the generation time, mobility or proximity of plants and animals concerned, and, for fisheries and stream biota, on the effect of reduced deposition on the frequency and intensity of acid episodes.

3.2.5 Current and future trends in acidification

3.2.5.1 Possible future changes
In considering factors which may affect surface water acidification in the future, it must be borne in mind that the inter-relationships between sulphur emissions, surface water acidity and its biological consequences are not linear.

The chemical reactions by which sulphur and nitrogen oxides are converted to acids, and the rate at which the acids are deposited, are affected by weather patterns. The interactions of the acids with soil and rock will depend on pollutant loading. There is strong evidence that episodic events of high acidity are important for freshwater ecosystems. For these reasons a change in emissions may not necessarily result in a uniform change in surface water acidification.

In September 1986 it was announced that CEGB will install flue gas desulphurisation (FGD) at three major power stations, the net effect of which will be to reduce sulphur emissions in the UK by about 14%. The first unit is expected to be operational in 1993. Since then there has been an agreement which commits the UK to reducing emissions from large combustion plants by 20% in 1993, 40% by 1998 and 60% by 2003.

Emission and deposition models in the UK show that about 70% of the deposition can be directly attributable to UK sources and that the remainder derives from the Atlantic or from Europe. It is conceivable that some of this latter deposition originates in the UK and returns as a result of changes in wind direction, but if the estimates of the models are correct then a 20% reduction in emission would result in a reduction in deposition of around 14% overall unless emissions from other sources were also reduced. No further improvements in the pH of surface waters are likely until after 1993, and they may well not be detectable until after the year 2000.

3.2.5.2 Industrial emissions
If no steps are taken to reduce emissions by changing the source of fuel, by using alternative energy sources or by applying control technology, then emissions will reflect energy consumption and industrial activity. Between 1970 and 1984 the recession caused emissions in the UK to fall but the economic revival has halted the decline, and

without remedial action sulphur emissions would probably show a slow rise. For the UK the indications are that energy consumption, and thus sulphur and nitrogen emissions, would probably increase only very slowly (see Table 3.2).

Table 3.2 UK sulphur emissions 1970–1987

	1970	1984	1985	1986	1987 (est.)
Sulphur emissions (million tonnes)	6·12	3·53	3·54	3·74	3·68

3.2.5.3 Vehicle emissions
Vehicle exhausts emit nitrogen oxides which may contribute to surface water acidity. The use of catalysts and lean-burn engines will reduce such emissions. On the basis of recently agreed EC emission limits, forecasts suggest that, despite a significant projected increase in traffic density, nitrogen oxide emissions could decrease by as much as 20%.

3.2.5.4 Land use
Current plans are to increase the area of afforestation. It is proposed that the bulk of this (90%) will occur in Scotland. There is evidence that afforestation increases the acidity of waters draining sensitive catchments in areas of acid deposition. Afforestation of such catchments can therefore be expected to exacerbate the effects of acid deposition and offset the benefits of a reduction in emissions.

Changes in subsidies are leading to the conversion of farmland of marginal productivity to forest. The area affected mainly include uplands in regions sensitive to acidification.

3.2.5.5 Climate
The 'greenhouse' effect, due to rising concentrations of carbon dioxide in the atmosphere, has been widely discussed. The evidence for change is still ambiguous but the global temperature has risen by 0·5–0·7°C over the past 100 years and models predict a temperature rise in the UK of $3 \pm 1·5$°C by 2050 (DoE Report).

The effects of such a change on world climate are difficult to predict but, to the extent that concentrations of acid in rainfall are dependent upon water patterns, any change in mean wind speed and direction or in the incidence of extreme rainfall events would be likely to alter acid deposition patterns. Longer dry spells followed by heavy rain could, for example, produce more intense acid episodes. A greater prevalence of easterly winds might increase the contribution from the Continent to UK deposition.

3.2.5.6 Models

In order to forecast reliably the likely effects of strategies to reduce acidification and its effects, mathematical models is required which can predict the consequences of changes in any of the factors which affect surface water acidification.

A range of models is available for assessing the impact of acid deposition and land-use change on stream water chemistry. These include models such as MAGIC (Model of Acidification of Groundwater in Catchments) for predicting long-term trends and the BIRKENES model, a daily simulation model developed originally for the Birkenes catchment in Norway. These models are particularly useful for understanding the complex interaction between physical and chemical behaviour. They have been applied to several UK catchments to predict short- and long-term responses to deposition reductions and land-use change, e.g. afforestation.

3.2.5.7 Future policy on emissions

Concern about the effects of emissions of sulphur and nitrogen oxide emissions has led to the formation of the '30% Club', a group of nations committed to reducing their emissions of sulphur by 30%. There is pressure for the UK to join the club but it must be pointed out that the effects of such a reduction will depend on how it is implemented. While the size of the total national reduction in sulphur emissions is important as far as long range transport of sulphur oxides is concerned, the maximum benefit to acidified UK surface waters would accrue if sites for the installation of FGD were selected from those whose emissions have the greatest influence on susceptible waters rather than from those at which the most cost effective reductions could be secured.

3.2.6 Acidification of waters and human health

Acidic precipitation will only have a significant effect on the pH of weakly buffered water and in general this implies soft water with a low bicarbonate content. A reduction of pH is of no direct concern in relation to human health although it could in theory lead to an increased concentration of metals such as iron, manganese and aluminium. These are presently of concern in water supply mainly from the aesthetic rather than the health point of view. If water having an acidic pH is allowed to pass into a water supply distribution system without any treatment then there may be opportunities for it to pick up further concentrations of iron and metals such as copper, zinc and lead that are common in household plumbing systems. Increased lead levels are of concern in relation to public health and the EC Directive on the Quality of Water Intended for Human Consumption sets a Maximum Acceptable Concentration of $50 \mu g$ litre^{-1}. In the United Kingdom most surface water supplies are filtered and their pH is adjusted prior to distribution to control corrosion in general, and plumbosolvency in particular. Problems would normally be expected in unfiltered water supplies and these would be worsened by a decrease in pH if this was not remedied.

The possibility of an involvement of aluminium in the etiology of Alzheimer's disease has received some attention recently but the evidence that aluminium may play a causative part in this disease is still equivocal. If the maximum admissable concentration of this metal in drinking water were reduced it is conceivable that the elevation in aluminium concentration which occurs in acidified waters might become significant for some small untreated water supplies. It would, however, be premature at present to regard water aluminium as a risk to health.

3.2.7 Management

3.2.7.1 Critical load assessment

Many countries are now accepting the need to reduce emissions of sulphur and nitrogen compounds but the best treatment option, in terms of cost/benefit, has yet to be determined.

In biological terms the best option would be to reduce loadings of sulphur and nitrogen compounds to a level which would lead to no harmful effects on biota. In many areas of the UK this would require a $> 50\%$ reduction in loadings.

In attempting to quantify critical loadings, the role of sulphur and nitrogen in acidifying soils and surface waters must be considered separately. Therefore, in the context of this section, only the effects on surface waters are considered.

The majority of sensitive upland catchments are nutrient-deficient and an increased supply of nutrients from the atmosphere would generally benefit primary production. The direct uptake of nitrogen compounds has a net acidifying effect on soils but will only cause surface water acidification if input exceeds utilisation with a consequent increase in stream concentrations. Nitrate levels in upland streams are generally $< 25 \mu$eq litre^{-1} except when tree harvesting occurs, when nitrate levels can increase two- to three-fold. A loading of 10–20 kg N ha^{-1} year^{-1} has produced increased nitrate levels in a few Swedish rivers and therefore the aim should be for loadings of < 10 kg ha^{-1} year^{-1} to protect the majority of sensitive catchments.

Sulphate is more conservative than nitrate and is considered to be less important in acidifying soils but more important in acidifying surface waters. Increasing sulphate concentrations result in decreased alkalinity (Acid Neutralising Capacity) and increased acidity, aluminium and base cation levels.

A positive alkalinity of at least 10–20 μeq litre^{-1} must be maintained at all times to protect biota. Using this criterion a sulphate loading of < 10 kg ha^{-1} year^{-1} is appropriate for the most sensitive catchments rising to 15 kg ha^{-1} year^{-1} for moderately sensitive catchments. It should be emphasised that these average loading values for nitrate and sulphate correspond to varying concentrations of these ions in precipitation, depending on rainfall amounts. Where catchments have plantations of semi-mature coniferous forests the loading should be even lower than the above targets, particularly where more than 50% of the area has been planted.

North-west Scotland is one of the few areas in the UK with a low pollutant loading, minimal surface water acidification and no reported biological problems. Using rain chemistry data from this area the target concentrations of excess sulphate in precipitation would be $< 25 \mu$eq litre^{-1} for moorland catchments and $< 15 \mu$eq litre^{-1} for forested catchments.

The target loading values used above are based on empirical data from lakes and soils in Scandinavia, USA and Canada plus additional information from process orientated models.

3.2.7.2 Liming techniques and costs

Various liming techniques have been used in Scandinavia and North America in an attempt to overcome surface water acidification, and these have recently been extended to parts of the UK. At Loch Dee in Galloway, scallop shells, limestone chips and powdered limestone have been added at different times to input streams, and limestone was also applied to parts of a stream catchment. Incidence of samples with pH < 5.0 at the loch outlet decreased following liming, although much of the calcium was taken up by sediment.

Direct application of powdered limestone to acid upland lakes has been carried out in West Wales (Underwood *et al.*, 1987). At Llyn Berwyn (surface area 13 ha; initial mean pH 4·3), water quality has been maintained at pH 5·5–7·0 since April 1985 by 8-t doses every 9–12 months. The mean annual cost is around £1000, and the previously extinct fishery now supports a stock of brown trout. The alternative approach of liming a catchment has been adopted at Loch Fleet in Galloway (surface area 17 ha; initial mean pH 4·4). At this site, a total dose of 350 t (up to 30 t ha^{-1} on parts of the catchment; mean dose 3·3 t ha^{-1} overall) has resulted in pH at the loch outlet rising to around 6·5. Catchment liming as practised at Loch Fleet has the advantage over direct lake liming of raising pH in the inlet streams, with possible implications for salmonid spawning. However, against this must be set the ecological damage to *Sphagnum*-dominated areas of the catchment arising from the high dosing rates adopted.

Application of limestone to a river catchment has been undertaken in the river Esk catchment in south-west Cumbria, where North West Water applied powdered limestone to c. 650 hectares (1600 acres) of agricultural land in the whole catchment at a rate of 4·9 tonnes/hectare (2 tonnes/acre) in October–November 1986. An immediate improvement in pH was noted which was maintained throughout the winter. A reduction in pH during spates was still apparent but not to the same extent as previously. The effects will continue to be monitored both chemically and biologically. Mortalities of fresh-run salmonids have previously occurred in the main river in 1980 and 1983, associated with acid flushes.

In conclusion, whilst the treatment of small lakes both by direct liming and by catchment liming has been shown to be feasible, the full ecological consequences of both methods have yet to be evaluated. Moreover, treating larger lakes would be prohibitively expensive, with Loch Dee, for example (surface area 100 ha; catchment area 1560 ha), costing some £10 000 per annum for direct application of limestone. Liming techniques are therefore best regarded as a temporary remedial

34

Air pollution, acid rain and the environment

measure to preserve vulnerable fisheries or to enable restoration of extinct fisheries in certain areas.

3.2.7.3 Afforestation policy

With economic pressures on upland agriculture likely to increase in the future due to policy reviews within the EEC, a continued demand for further afforestation in areas vulnerable to acidification seems likely. A recent projection has estimated that the total area of afforested land may treble by the end of the century.

Clear evidence exists to demonstrate a link between afforestation and surface water acidity in several areas (Section 3.2.3), although the relative importance of atmospheric inputs and drainage effects in accounting for this is poorly understood at present. Investigations into the relationship between acid deposition, forest management and stream acidity are currently underway in both Wales and Scotland. The results of these studies in the next 2–3 years will be important in determining what measures, if any, may be effective in minimising adverse impacts on the aquatic environment. In the meantime revised guidelines for forestry interests in the Welsh uplands are being formulated by the Welsh Water Authority and the Institute of Terrestrial Ecology, and it is hoped that these guidelines will be extended nationally.

No statutory planning framework exists at present to control forestry development, although consultations with local authorities are carried out by the Forestry Commission in connection with grants to private forestry schemes. This consultative procedure was extended in Wales in 1984 to include the water authority, out of a concern to overcome the problems arising from acidification. Whilst most schemes have been approved either as originally proposed or after agreed amendment, outright objection to one scheme was made in a particularly sensitive river catchment. After further consultation the Forestry Commission recommended that the proposed development should not receive grant aid, on the grounds that a detrimental effect on river quality following afforestation was likely. A similar consultation procedure has been agreed with the Clyde River Purification Board and would be of value in all areas vulnerable to acidification. However, to be effective this would need to apply to all new forestry, and not merely to grant-aided schemes.

3.3 CONCLUSIONS

3.3.1 Cause and extent of acidification

It is now clear that many surface waters in acid-sensitive areas of the UK have been acidified, and that the primary cause of acidification has been acid deposition, confirming the cause–effect relationships first proposed in Scandinavia and Canada. However, although a considerable amount of work is in progress, it is still not known how much of the United Kingdom is acidified (cf. Fig. 3.1). So far it has been established that recent acidification (post-1850) has occurred in Galloway, Rannoch Moor and Arran in Scotland, in the Tywi headwaters and the Rhinogs of Wales, and in some of the high tarns in the English Lake District. Extrapolating from these results on the basis of geology and water chemistry it is also highly likely that many of the streams and lakes in the Cairngorms, streams and reservoirs in the English Pennines, and streams in the heathland areas of south-east England are strongly acidified (Fig. 3.1).

3.3.2 Biotic effects

In most cases acidification has taken place over the last 130 years or so, but sediment core records suggest that acute acidification to pH values less than 5 occurred in the period 1930–1970. The long history of acidification in the UK as compared to other countries is to be expected in view of the United Kingdom's history of heavy industrialisation, and many unrecorded biotic changes undoubtedly took place prior to the scientific interest in acid rain and its effects during the last decade. Nevertheless, sufficiently good records exist to document the decline of brown trout fisheries in Galloway and Wales and natterjack toad populations in the South-east of England, and core data can be used to show algal and invertebrate change throughout the country.

3.3.3 Afforestation

Recent research has shown that conifer afforestation can also be an acidifying influence, exacerbating the problem at sites already partially acidified and causing acidification at others. It is not yet clear how much this acidification is due to the direct effect of tree growth, and how much to the interaction of the growing forest canopy with a polluted atmosphere, but the main effect on water quality occurs after the time of canopy closure, some 10–15 years after planting. Because of this it is predicted that further acidification of surface waters will occur as young forest matures.

3.3.4 Reversibility

Because of sulphur stores in the soil there may be an initially slow response to a decline in acid deposition, but there are already signs that a small improvement has occurred in non-afforested Galloway lakes following the post-1979 emission (deposition?) decline (Battarbee *et al.*, 1988). However, this is counter to the changes in lakes with afforested catchments where further acidification has taken place since 1970. If the forest effect is primarily through scavenging, this implies that an increase in the rate of emission reduction is required to prevent acidification of waters where there are young plantations or where new forests are planned. In the absence of such a reduction in the next decade, liming forest soils prior to canopy closure may help to neutralise throughflow and stemflow acidity, and prevent the transfer of acidity to surface waters.

RECOMMENDATIONS

(i) The results of the baseline survey should be used to establish long-term monitoring stations at key localities.
(ii) Research into the mechanisms of forest effects should be continued and expanded as a priority in view of the substantial expansion of afforestation now proposed.
(iii) Research into wildlife implications should be regarded as a continuing priority, for example, by the Natural Environmental Research Council.
(iv) Liming of forest catchments should be used where appropriate as a short-term prophylactic.
(v) The closer liaison between the Forestry Commission and water authorities now evident should be speeded up and the potential impact of new afforestation schemes on surface waters should be always taken into account to ensure that acidification is prevented or alleviated.
(vi) The need for statutory control of planning applications for afforestation should be considered because land is now being planted without grant aid, thus bypassing Forestry Commission controls.

REFERENCES

ANDERSON, N. J., BATTARBEE, R. W., APPLEBY, P. G., STEVENSON, A. C., OLDFIELD, F., DARLEY, J. & GLOBER, G. (1986) Palaeolimnological evidence for the recent acidification of Loch Fleet, Galloway. *Palaeoecology Research Unit, University College London, Research Papers,* No. 17.

BARRETT, C. F., ATKINS, D. H. F., CAPE, J. N., FOWLER, D., IRWIN, J. G., KALLEND, A. S., MARTIN, A., PITMAN, J. I., SCRIVEN, R. A. & TUCK, A. F. (1983). *Acid Depositions in the United Kingdom.* Warren Spring Laboratory, Stevenage.

BATTARBEE, R. W. (1984). Diatom analysis and the acidification of lakes. *Phil. Trans. R. Soc.,* **B305**, 451–77.

BATTARBEE, R. W. & CHARLES, D. F. (1986). Diatom-based pH reconstruction studies of acid lakes in Europe and North America: a synthesis. *Water, Air & Soil Pollut.,* **30**, 347–54.

BATTARBEE, R. W., FLOWER, R. J., STEVENSON, A. C. & RIPPEY, B. (1985). Lake acidification in Galloway: a palaeoecological test of competing hypotheses. *Nature Lond.,* **314**, 350–2.

BATTARBEE, R. W., FLOWER, R. J., STEPHENSON, A. C., JONES, V. J., HARRIMAN, R. & APPLEBY, P. G., (1988). Diatom and chemical evidence for reversibility of acidification of Scottish lochs. *Nature Lond.,* **332**, 530–2.

BEEBEE, T. J. C. (1976). The natterjack toad *Bufo calamita* in the British Isles: a study of past and present status. *Brit. J. Herpetol.,* **5**, 515–21.

BEEBEE, T. J. C. (1977). Environmental change as a cause of natterjack toad *Bufo calamita* declines in Britain. *Biol. Conserv.,* **11**, 87–102.

BEEBEE, T. J. C. (1986). Acid tolerance of natterjack toad *Bufo calamita* development. *Herpetol. J.,* **1**, 78–81.

BEEBEE, T. J. C. & GRIFFIN, J. R. (1977). A preliminary investigation into natterjack toad *Bufo calamita* breeding site characteristics in Britain. *J. Zool.,* **181**, 341–50.

BURNS, J. C., COY, J. S., TERVET, D. J., HARRIMAN, R., MORRISON, B. R. S. & QUINE, C. P. (1984). The Loch Dee project: a study of the ecological effects of acid precipitation and forest management on an upland catchment in south-west Scotland, 1. Preliminary investigations. *Fish. Management,* **15**, 145–67.

CUMMINS, C. P. (1986). Effects of aluminium and low pH on growth and development in *Rana temporaria* tadpoles. *Oecologia,* **69**, 248–52.

DONALD, A. P. & STONER, J. H. (in press). The quality of atmospheric deposition in Wales. *Archives of Environmental Contamination and Toxicology.*

EDMUNDS, W. M. & KINNIBURGH, D. G. (1986). The susceptibility of UK groundwaters to acidic deposition. *J. Geol. Soc. Lond.,* **143**, 707–20.

FLOWER, R. J. & BATTARBEE, R. W. (1983). Diatom evidence for the recent acidification of two Scottish lochs. *Nature Lond.,* **305**, 130–3.

FLOWER, R. J., BATTARBEE, R. W. & APPLEBY, P. G. (1987). Palaeolimnological studies in Galloway: lake acidification and the role of afforestation. *J. Ecol.*

Freshwater Fisheries Laboratory (1982–84). *Triennial Review of Research.* Department of Agriculture and Fisheries for Scotland.

FRITZ, S. C., STEVENSON, A. C., PATRICK, S. T., APPLEBY, P. G., OLDFIELD, F., RIPPEY, B., DARLEY, J. & BATTARBEE, R. W. (1986). Palaeoecological evaluation of the recent acidification of Welsh lakes, 1. Llyn Hir, Dyfed. *Palaeoecology Research Unit, University College London, Research Papers,* No. 16.

HARRIMAN, R. & MORRISON, B. R. S. (1982). The ecology of streams draining forested and non-forested catchments in an area of central Scotland subject to acid precipitation. *Hydrobiologia*, **88**, 251–63.

HARRIMAN, R. & WELLS, D. E. (1985). Causes and effects of surface water acidification in Scotland. *J. Water Pollut. Control*, **84**, 3–10.

HARRIMAN, R., MORRISON, B. R. S., CAINES, L. A., COLLEN, P. & WATT, A. W. (1987). Long-term changes in fish populations of acid streams and lochs in Galloway, South-west Scotland. *Water, Air & Soil Pollut.*, **32**, 89–112.

HOWELLS, G. D. & BROWN, D. J. A. (1986). Loch Fleet: techniques for acidity mitigation. *Water, Air & Soil Pollut.*, **30**, 817–26.

JONES, V. J., STEVENSON, A. C. & BATTARBEE, R. W. (1986). Lake acidification and the land-use hypothesis: a mid-post-glacial analogue. *Nature Lond.*, **322**, 157–8.

KREISER, A., STEVENSON, A. C., PATRICK, S. T., APPLEBY, P. G., RIPPEY, B., DARLEY, J. & BATTARBEE, R. W. (1986). Palaeoecological evaluation of the recent acidification of Welsh lakes, II. Llyn Berwyn, Dyfed. *Palaeoecology Research Unit, University College London, Research Papers*, No. 18.

KRUG, E. C. & FRINK, C. R. (1983). Acid rain on acid soil: a new perspective. *Science, N.Y.*, **221**, 520–5.

MAITLAND, P. S., LYLE, A. A. & CAMPBELL, R. N. B. (1987). *The status of fish populations in waters vulnerable to acid deposition in Scotland*. Institute of Terrestrial Ecology, Grange-over-Sands.

North-West Water (1986). *Final report to European Commission Contract no. 867 UK(H)*. North-West Water Authority, Warrington.

ORMEROD, S. J. & EDWARDS, R. W. (1985). Stream acidity in some areas of Wales in relation to historical trends in afforestation and the usage of agricultural limestone. *J. Environ. Manage.*, **20**, 189–97.

ORMEROD, S. J. & TYLER, S. J. (1987). Dippers *Cinclus cinclus* and grey wagtails *Motacilla cinerea* as indicators of stream acidity in upland Wales. In *Birds as bio-indicators*, ed. A. Diamond, *ICBP Tech. Publs*.

ORMEROD, S. J., ALLINSON, N., HUDSON, D. & TYLER, S. J. (1986). The distribution of breeding dippers *Cinclus cinclus* (L.)(Aves) in relation to stream acidity in upland Wales. *Freshwater Biol.*, **16**, 501–7.

ORMEROD, S. J., BOOLE, P., McCAHON, C. P., WEATHERLEY, N. S., PASCOE, D. & EDWARDS, R. W. (1987). Short-term experimental acidification of a Welsh stream: comparing the biological effects of hydrogen ions and aluminium. *Freshwater Biol.*, **17**, 341–56..

PATRICK, S. T. (1987). Palaeoecological evaluation of the recent acidification of Welsh lakes. *Palaeoecology Research Unit, University College London, Research Papers*, No. 21.

PATRICK, S. T. & STEVENSON, A. C. (1986). Palaeoecological evaluation of the recent acidification of Welsh lakes, III. Llyn Conwy & Llyn Gamallt, Gwynedd. *Palaeoecology Research Unit, University College London, Research Papers*, No. 19.

PENNINGTON, W. (1984). Long-term natural acidification of upland sites in Cumbria: evidence from post-glacial lake sediments. *Rep. Freshwat. Biol. Ass.*, **52**, 28–46.

PENNINGTON, W., HAWORTH, E. Y., BONNY, A. P. & LISHMAN, J. P. (1972). Lake sediments in northern Scotland. *Phil. Trans. R. Soc.*, **B**, **264**, 191–294.

RAVEN, P. J. (1985). The use of aquatic macrophytes to assess water quality changes in some Galloway lochs: an exploratory study. *Palaeoecology Research Unit, University College London, Research Papers*, No. 9.

RAVEN, P. J. (1986). Occurrence of *Sphagnum* moss in the sublittoral of several Galloway lochs, with particular reference to Loch Fleet. *Palaeoecology Research Unit, University College London, Research Papers*, No. 13.

ROSENQUIST, I. T. (1978). Alternative sources for acidification of river water in Norway. *Sci. Total Environ.*, **10**, 39–49.

ROUND, F. E. (1957). The late-glacial and post-glacial diatom succession in the Kentmere Valley deposit, 1. Introduction, methods and flora. *New Phytol.*, **56**, 98–126.

STEVENSON, A. C., PATRICK, S. T., FRITZ, S. C., APPLEBY, P. G., RIPPEY, B., OLDFIELD, F., DARLEY, J., HIGGITT, S. R. & BATTARBEE, R. W. (1987). Palaeoecology evaluation of the recent acidification of Welsh lakes, IV. Llyn Gynon, Dyfed. *Palaeoecology Research Unit, University College London, Research Papers*, No. 20.

STONER, J. H. & GEE, A. S. (1985). Effects of forestry on water quality and fish in Welsh rivers and lakes. *J. Inst. Water Engnrs & Scient.*, **39**, 27–45.

STONER, J. H., GEE, A. S. & WADE, K. R. (1984). The effects of acidification on the ecology of streams in the Upper Tywi catchment in West Wales. *Environ. Pollut. Ser. A*, **35**, 125–57.

SUTCLIFFE, D. W. (1983). Acid precipitation and its effects on aquatic systems in the English Lake District (Cumbria). *Rep. Freshwat. Biol. Ass.*, 51st, 30–62.

SUTCLIFFE, D. W. & CARRICK, T. R. (1986). Effects of acid rain on waterbodies in Cumbria. In *Pollution in Cumbria*, ed. by P. Ineson. Institute of Terrestrial Ecology, Abbots Ripton, pp. 16–25.

SUTCLIFFE, D. W., CARRICK, T. R., HERON, J., RIGG, E., TALLING, J. F., WOOF, C. & LUND, J. W. G. (1982). Long-term and seasonal changes in the chemical composition of precipitation and surface waters of lakes and tarns in the English Lake District. *Freshwater Biol.*, **12**, 451–506.

UNDERWOOD, J., DONALD, A. P. & STONER, J. H. (1987). Investigations into the use of limestone to combat acidification in two lakes in west Wales. *J. Environ. Manage.*, **24**, 29–40.

Watt Committee (1984). *Report No. 14—Acid Rain*. The Watt Committee on Energy, London.

WEST, G. (1910) A further contribution to a comparative study of the dominant phanerogamic and higher cryptogamic flora of aquatic habitat in Scottish lakes. *Proc. R. Soc. Edinb.*, **30**, 65–81.

Section 4

Corrosion of Building Materials due to Atmospheric Pollution in the United Kingdom

Michael Manning

Central Electricity Research Laboratories, CEGB, Leatherhead, Surrey

This paper presents the work of a sub-group of the
Watt Committee working group on Air Pollution,
Acid Rain and the Environment.

Membership of Sub-group

Dr G. B. Gibbs } (Joint Chairmen)
Dr M. I. Manning }

J. Bernie
Dr R. N. Butlin
Prof. R. U. Cooke
Dr W. M. Edmunds
Dr J. E. Harris
Dr J. B. Johnson
Dr G. Lloyd
Prof. K. Mellanby
Dr Clifford Price
Dr M. L. Williams

4.1 INTRODUCTION

The lifetime for which buildings are designed can vary enormously. Secular buildings designed for a specific purpose tend to become obsolescent because that purpose disappears, and the structure is too inflexible to enable it to be adapted for modern purposes. Such obsolescence generally occurs within a period somewhat shorter than a century. If such buildings have been well maintained, corrosion of the fabric due to attack from the atmosphere is rarely life-determining (though there have been spectacular exceptions to this rule in the post-war period). On the other hand, ecclesiastical buildings, monuments and works of art have, notionally at least, a design life of infinity.

Atmospheric pollution can accelerate the degradation of inorganic building materials and, whilst this may shorten the life of certain structures, it more generally means that additional costs have to be met for structural maintenance. Many historic buildings in the UK are constructed of limestone and the statuary and detailing is sometimes thought of as being irreplaceable. This is a rather shortsighted viewpoint, as limestone would have a finite durability even in the absence of any man-made atmospheric pollutants. Consequently, it is only prudent to plan for the maintenance and eventual replacement of cultural artefacts.

The rate of materials degradation and the corresponding maintenance requirements depend, among other things, on the levels of urban pollutants. This report describes the trends in pollutant emissions and the resulting trends in urban air quality. The pathways for pollutant deposition onto materials surfaces are then considered. Pollutants can arrive in rain or else be presented to the surface by gaseous deposition onto moist surfaces. The rates of degradation are considered, first, for metals and then for stone and concrete, etc.

Concentrations of sulphur dioxide, probably the most damaging pollutant, have declined markedly in recent decades but, in spite of this, high rates of materials damage are still reported. For porous materials like stone, the surface may have been permanently altered by its past exposure to high levels of sulphur dioxide and the present high rates of damage indicate a 'memory' of the previous exposure history. Even for the case of zinc, where several studies have shown that the damage rate is directly proportional to SO_2 concentration, there is an indication that damage rates have not fallen over the years in proportion to the SO_2 reductions. It is

becoming increasingly likely that there are important synergisms between pollutants, especially between nitrogen oxides (NO_x) and SO_2 levels. Urban NO_x levels have not shown the same reduction in recent decades and so may now be relatively more important.

It is hard to attain a realistic perspective about the importance of present-day levels of environmental pollutants in influencing buildings maintenance costs. For stone, the relative importance of physical processes like freezing and salt action, and chemical processes due to pollution, are hard to assess. Nowadays stone damage rates seem to be less than proportional to SO_2 level, for example. For metals, the effects of salts derived from the sea have always been important, especially for steel. Marine and climatic influences mean, therefore, that the UK would always be a fairly aggressive place for many materials.

The most significant cause of accelerated buildings maintenance costs has been the corrosion of steel reinforcement in concrete structures. There are almost no data to relate the rate of reinforcement corrosion to the prevailing atmospheric conditions. If, however, atmospheric pollution had even a small influence on the corrosion of reinforced concrete, the economic significance of such an effect would dwarf most of the other forms of buildings degradation that have received more attention.

4.2 TRENDS IN UK EMISSIONS AND AIR QUALITY

The most important pollutants that affect structural materials are sulphur dioxide and oxides of nitrogen and their oxidation products, together with chlorides and particulate matter. This section describes trends in the emissions, air quality and deposition in the UK of those pollutants for which sufficient information is available. Ozone has also been considered and recent air quality data for this species will also be discussed.

4.2.1 Sulphur dioxide

There is a considerable amount of information available on the emissions, concentrations and depositions of sulphur species and these will therefore be discussed separately.

4.2.1.1 UK emissions of SO_2
Sulphur dioxide emissions arising from man's activities are calculated annually by Warren Spring

Laboratory (WSL) and are published in the annual DoE Digest of Environmental Protection and Water Statistics. Emissions have also been calculated for the period 1853 to 1974 by Bettelheim & Littler (1979). Their estimates are compared with those of WSL in Fig. 4.1. The two sets of calculations are based on similar figures for fuel use; the differences between the two sets of figures, which amount to 5–10%, arise chiefly from uncertainties over sulphur contents of fuels, from different assumptions regarding the amounts of sulphur retained in the ash following coal combustion and from the fuel use categories included by the two calculations. Bearing in mind that the emissions are calculated from fuel deliveries rather than use for many categories, the overall confidence in the estimates is probably 15–20% for the recent estimates (since 1950) but the earlier data must be considered more uncertain.

Both sets of estimates yield similar trends, rising steadily from a total of 1·3 million tonnes in 1850 to a peak in the decade 1960–1970. From the peak of 6·1–6·6 million tonnes in 1970, UK emissions have now decreased to levels similar to those in 1910.

The reduction in SO_2 emissions since 1980 has been due in broadly equal proportions to energy economies, reductions in sulphur contents of fuels, changes in fuel use patterns (e.g. to natural gas) and industrial modernisation and decline.

In terms of the proportional contribution to SO_2 emissions, significant changes have taken place since 1950. At that time sources emitting at low and medium rates such as domestic, commercial and industrial users were responsible for some 80% of SO_2 emissions while by 1970 this proportion was 50% (Weatherley *et al.*, 1975). By 1984 the proportional contribution from low and medium level emissions had decreased further to 25%, with power stations now contributing some 70%. A graph of source contributions to emissions of SO_2 is shown in Fig. 4.2. These trends have important consequences for urban (and rural) air quality in terms of both absolute concentrations and the proportional contribution from the major source categories.

4.2.1.2 SO_2 concentrations

Following the London smogs of the early 1950s and the Clean Air Act of 1956, a continuing programme of smoke control was introduced in the UK. In order to provide comparable data of a consistent minimum quality for the wide range of areas (urban/industrial/rural) found in the UK, the National Survey of Smoke and Sulphur Dioxide was set up in 1961. The monitoring sites initially numbered about 500, increasing to about 1200 by 1966. They were equipped and operated by the local authorities and other co-operating organisations and the Survey was coordinated by WSL, who processed, analysed and published the data. The concentrations of SO_2 and smoke in urban areas have decreased greatly since the start of the Survey, and the number of sites in operation has also decreased. Currently some 550 sites are in operation.

The reduction in annual average SO_2 concentrations averaged over all urban sites in the national survey is shown in Fig. 4.2. The decrease in concentrations parallels the decrease in emissions from the low and medium level sources, as is to be expected in qualitative terms since the high level emissions make a proportionally much lower contribution to ground level concentrations. A note of caution against over-generalisation is necessary at this stage, however. The precise contribution from

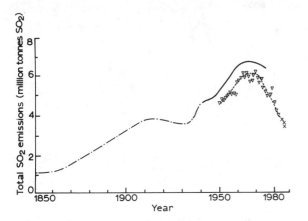

Fig. 4.1. Sulphur dioxide emissions in the United Kingdom from 1850: - - -, ———, Bettelheim & Littler (1979); — — ▽ — —, Warren Spring Laboratory.

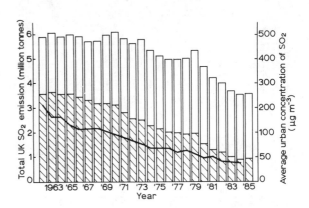

Fig. 4.2. Trends in SO_2 emissions and air quality. ———, Concentration; unshaded areas denote high level sources, shaded areas denote other sources.

different source categories will obviously vary from one urban area to another depending on the source mix, and can only be quantified by a detailed study of each individual area. Nonetheless, over the past thirty years or so the large decreases in urban SO_2 concentrations have clearly arisen primarily from the decrease in domestic and industrial/commercial emissions.

The foregoing discussion presents the national picture, but in order to give some indication of the changes in exposure to SO_2 of buildings and locations of interest, some time series plots for individual sites have been analysed. Two of the longest time series for airborne SO_2 concentrations which exist in the UK are those for County Hall, Westminster (OS Grid reference TQ307797) and for the site at Surrey Street in Sheffield (SK354871, identified as Sheffield 2 in the National Survey). Virtually complete records exist for both sites from 1932/33. Results from the Sheffield site, obtained using peroxide bubblers, are shown in Fig. 4.3. In both cases SO_2 concentrations remained broadly constant between 1930 and 1960 at 300–400 $\mu g\,m^{-3}$. Measurements ceased at Sheffield 2 in 1977 but annual mean concentrations in central Sheffield are currently of the order of 50–60 $\mu g\,m^{-3}$.

Similar time series have been plotted for sites close to buildings of interest, although the time series are shorter than that shown in Fig. 4.3. Data for the two sites in the National Survey closest to St Paul's Cathedral are shown in Fig. 4.4 where, as well as annual means, the 98th percentile of daily average values over a year are plotted as a reasonable measure of peak SO_2 concentrations.

Concentrations at the two sites in York closest to the Minster, namely York 5 (0·9 km distant) and York 3 (0·5 km distant), decreased, respectively, from annual averages of 116 $\mu g\,m^{-3}$ in 1963/64 and

155 $\mu g\,m^{-3}$ in 1962/63 to 70 $\mu g\,m^{-3}$ in 1979/80 and 74 $\mu g\,m^{-3}$ in 1971/72, when the site measurements ceased. Current annual mean concentrations at York 8 (0·9 km distant from the Minster) are 40–50 $\mu g\,m^{-3}$ and at York 6 (0·5 km distant), 40 $\mu g\,m^{-3}$. Peak daily mean concentrations as measured by the 98th percentiles over a year have similarly decreased since the early 1960s, for example, from 600 $\mu g\,m^{-3}$ to 200 $\mu g\,m^{-3}$ at York 5 (1963/64 to 1979/80); current 98th percentiles of daily mean SO_2 concentrations at York sites are 100–120 $\mu g\,m^{-3}$.

The longest runs of data available for sites in Lincoln are for the Lincoln 5 site (0·7 km south-west of the Cathedral) and for the sites Lincoln 11 and 12 which are on the southern edge of the city, 1·5–2 km distant from the Cathedral. At Lincoln 5 annual average SO_2 concentrations have decreased from 152 $\mu g\,m^{-3}$ in 1961/62 to 38 $\mu g\,m^{-3}$ in 1984/85; at Lincoln 11 concentrations decreased from 169 $\mu g\,m^{-3}$ in 1962/63 to 47 $\mu g\,m^{-3}$ in 1981/82 with a decrease from 133 $\mu g\,m^{-3}$ to 40 $\mu g\,m^{-3}$ over the same period at Lincoln 12. The 98th percentiles at the Lincoln 11 and 12 sites in 1962/63 were 557 and 410 $\mu g\,m^{-3}$, respectively, while in 1981/82 the range of 98th percentiles of SO_2 at the three sites was 97–137 $\mu g\,m^{-3}$.

It is worth noting here that the above discussion of peak concentrations has used the statistic of the 98th percentile of daily means over a year as the measure; maximum daily concentrations will, of course, vary more from year to year.

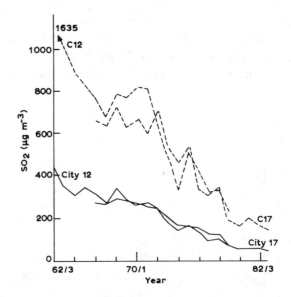

Fig. 4.4. SO_2 concentrations at two sites close to St Paul's Cathedral showing annual means (——) and indicating much higher values recorded on some days (98th percentile of daily means (----)).

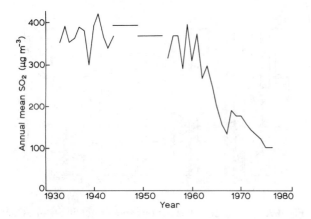

Fig. 4.3. Annual average SO_2 concentrations at a site in Sheffield (Sheffield 2).

Information on urban SO_2 concentrations before the 1930s is scarce but an analysis of coal imports into London from 1600 (Brimblecombe, 1977) suggests that coal use per unit area was broadly constant from 1800–1900, although Brimblecombe's analysis suggests a halving of SO_2 concentrations from 1900–1950. The reasons for this may lie in the model used; the measured data suggest that SO_2 annual averages stayed broadly constant from 1930–1960, which is probably due to increased oil consumption. Nonetheless, it seems likely that SO_2 concentrations in London during the nineteenth century were at least as high as those measured in the 1930s, i.e. 300–400 $\mu g\,m^{-3}$ annual average. Using a ratio of 2·8 as a typical value of 98th percentile/annual mean for SO_2 (obtained by WSL from an analysis of National Survey data) suggests that 98th percentiles of 850–1200 $\mu g\,m^{-3}$ were likely in London during this period; peak values in years with more extreme winters than average could have been higher than this.

As an indication of the peak concentrations reached during 'smog' episodes, during the December 1952 episode daily average concentrations of 3000–4000 $\mu g\,m^{-3}$ were measured at some sites in London. For a more detailed discussion the reader is referred to a report on a high pollution period in December 1975 (Keddie *et al.*, 1977), when daily mean SO_2 concentrations of up to 1200 $\mu g\,m^{-3}$ were observed at some sites in London.

4.2.2 Smoke emissions

As is the case for SO_2, detailed estimates of smoke emissions (from coal combustion) have been calculated from 1960. These are shown in Fig. 4.5, from which it is clear that there has been a major

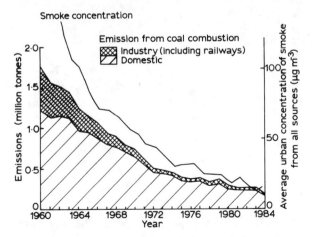

Fig. 4.5. Trends in smoke emissions and air quality.

decrease in overall emissions of smoke from coal combustion, especially in the domestic sector, as areas under smoke control have increased. This decrease in emissions has had a marked effect on urban black smoke concentrations, the trend in which is also shown in Fig. 4.5. Annual mean urban smoke concentrations in the early 1960s were about 140 $\mu g\,m^{-3}$ with some areas reporting concentrations of up to 350 $\mu g\,m^{-3}$ annual average. In London, for example, the urban average annual mean smoke concentration was 150 $\mu g\,m^{-3}$ in the late 1950s and an estimate of 300 $\mu g\,m^{-3}$ was made for 1940–50 (WSL, 1972). Estimates for historical smoke concentrations in London have been made by Brimblecombe (1977). He quoted concentrations averaged over the whole of London, so the absolute values may not be strictly comparable with the values quoted above; however, this information on trends suggests that smoke concentrations peaked in 1900 and declined by about 50% up to 1930–40. Smoke concentrations in London during the nineteenth century were as high as those in the mid-twentieth century and could well have been up to a factor of two higher towards the late 1800s.

In terms of peak concentrations, a typical ratio of 98th percentile of daily means over a year to annual mean for smoke concentrations in urban areas in the UK is 3·8. In the late 1950s to early 1960s, 98th percentile smoke concentrations were typically 550–1300 $\mu g\,m^{-3}$. During the 'smog' episodes referred to earlier, peak daily concentrations of smoke reached 4000 $\mu g\,m^{-3}$ at two sites on one day in London in December 1962. (Smoke concentrations were not available during the 1952 episode but since it was estimated (WSL, 1967) that smoke emissions halved between 1952 and 1962, smoke concentrations during the 1952 episode could have been up to twice those observed during the December 1962 episode —the meteorological conditions were very similar.)

Smoke concentrations at present are very much lower than they were before the mid-1950s. The UK average of urban annual means is now 17 $\mu g\,m^{-3}$. Smoke annual mean concentrations in Central London are now typically 25–35 $\mu g\,m^{-3}$ while those in York and Lincoln are 15–20 $\mu g\,m^{-3}$ and 15 $\mu g\,m^{-3}$, respectively. This has the consequence that in such areas motor—and, in particular, diesel —vehicles are often the predominant contributor to smoke concentrations. In Central London, for example, it is likely that up to 80% of the black smoke annual average concentration arises from motor vehicles (Ball & Hume, 1983).

4.2.3 Oxides of nitrogen

Nitrogen oxides are produced by combustion processes using air as oxidant when the combustion temperature is 1800°C or higher, and to some extent from nitrogen-containing materials in fossil fuels. Total UK emissions of nitrogen oxides were in the region 1·5–2·0 million tonnes in 1980, having approximately doubled since 1945. There has been some decline in recent years because of reductions in total fuel usage, but this does not seem to have been reflected in urban NO_x levels. About 37% of the total nitrogen oxides arise from power stations and 47% from motor vehicles; the latter contribute greatly to the concentration in air at ground level in cities. Nitrogen dioxide levels are typically about $10\,\mu g\,m^{-3}$ in rural areas, $40–70\,\mu g\,m^{-3}$ in towns away from roads, and as much as $80–100\,\mu g\,m^{-3}$ at the roadside (annual median value). Hourly values may exceed $200\,\mu g\,m^{-3}$ in London. More complex considerations probably apply to deposition of nitrogen oxides compared with sulphur dioxide. Broadly 30% of the acidity of rainwater may be due to nitric acid in Scotland and 60% in East Midlands (Barrett *et al.*, 1983). It is not known what proportion applies in cities, but with the large numbers of motor vehicles, it is likely that the second figure applies better here.

Estimates of UK emissions of NO_x from 1905 to 1980 have been made by WSL (Barrett *et al.*, 1983) and are shown in Fig. 4.6 disaggregated by fuel type. The broad trend is one of roughly constant emissions up until 1945, after which emissions increased by approximately a factor of two to their present levels. Unlike SO_2, however, there is less indication that NO_x emissions peaked in the late 1960s/early 1970s. Yearly estimates of NO_x emissions have been published by Department of the Environment and these are also shown in Fig. 4.6. However, the method of calculating motor vehicle emissions has recently been improved based on measurements made at WSL (Rogers, 1984), which resulted in the estimates of NO_x emissions for 1983 being revised upwards by 10%. Estimates for subsequent years will use the new method but a revision of earlier years is difficult to carry out accurately as the WSL measurements underpinning the new method relate to the UK car fleet in 1981–83. However, approximate revisions for earlier years have been carried out and these rather than the DoE Digest totals are shown in Fig. 4.6. The recent data from 1974 show no clear trend in national NO_x emissions either in total NO_x or in the

emissions from low- and medium-level sources (i.e. excluding power station and refinery emissions), although there is an indication that total NO_x emissions decreased between 1979 and 1984 largely due to a decrease in this period of some $250\,kt\,year^{-1}$ in power station emissions. 1984 was possibly a slightly anomalous year because of the dispute in the coal industry, so that caution is necessary in drawing firm conclusions from these figures.

Air concentration data for nitrogen oxides in the UK are very much less plentiful than is the case for SO_2. The longest run of measurements at an urban location is that for the WSL Central London Laboratory in Victoria and a time series plot of NO_x concentrations is shown in Fig. 4.7. The plot shows monthly average concentrations and the 99th percentile of hourly average concentrations in each month. Any significant trend in annual emission rates would be detected in the longer averaging time concentrations (monthly to yearly averages), as is clearly shown in the cases of smoke and sulphur dioxide. However, in the case of NO_x there is no significant trend in the monthly or annual average concentrations. (The overall accuracy and precision of the measurements is 10%.) Peak hourly concentrations such as the 99th percentiles plotted here are determined more by variations in meteorologi-

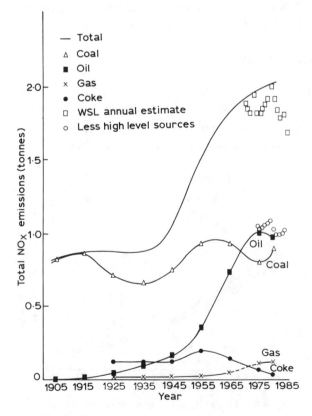

Fig. 4.6. NO_x emissions in the UK from 1905.

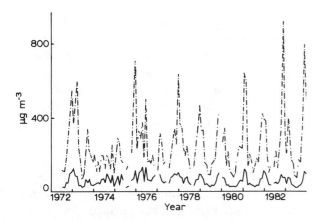

Fig. 4.7. NO_x concentrations in Central London. Monthly average concentrations (——) and 99th percentile of hourly average concentrations in each month (–·–·–).

cal conditions than by emissions and tend to be a less satisfactory indicator of trends than longer term averages. Where major emission trends have taken place, however, even short-term concentrations mirror the trends (see, for example, the 98th percentiles of smoke and SO_2 daily means over a year discussed above) but in the case of NO_x in Fig. 4.7 no such trends are apparent over the 11 years of data plotted.

4.2.4 Ozone

As with NO_x, measurements of ozone over extended periods are scarce in the UK. Unlike the pollutants considered so far ozone is not a primary pollutant. There are no significant man-made sources of ozone in the troposphere or atmospheric boundary layer (the lowest 1–2 km or so of the atmosphere). Ozone is produced in high concentrations in the stratosphere by UV irradiation and is thence transported in the free troposphere, supplementing ozone produced photochemically within the free troposphere. These 'background' tropospheric concentrations can be enhanced in the atmospheric boundary layer by elevated ozone concentrations produced in so-called 'photochemical episodes' by the action of sunlight on ozone-generating precursor pollutant emissions of nitrogen oxides and reactive volatile hydrocarbon/organic compounds. For a more detailed discussion of atmospheric ozone the reader is referred to a recent review (Derwent, 1986a, b).

The longest set of ozone measurements in an urban area in the UK is that obtained by WSL at the Central London Laboratory, and a time series of monthly average concentrations and 99th percentile of hourly averages in each month is shown in

Fig. 4.8. There is considerable variation from year to year in the 99th percentile concentrations but no clear trend is discernible in the data. The comments made in section 4.2.3 above on the predominant role of short-term meteorological variations in determining short period (in this case, hourly) concentrations are even more relevant for the secondary pollutant ozone where not only are poor dispersion conditions necessary for elevated ozone concentrations to be produced but also sufficient sunlight has to be present, together with favourable air mass trajectories transporting precursor pollutants from source to receptor.

4.2.5 Chloride

There have been no systematic measurements of chloride or HCl over an extended period in the ambient atmosphere in the UK, although the concentrations of Cl^- in rainwater have been measured by a number of organisations over a period of years. These latter measurements tend to be dominated by marine-derived chloride and year-to-year variations in the measured concentrations are generally reflections of meteorological differences rather than trends in pollutant emissions. Annual precipitation weighted mean chloride (Cl^-) concentrations in rainfall at Goonhilly near the Cornish coast, for example, from 1981–85 were 290–470 μeq litre^{-1}, whereas the corresponding non-marine sulphate concentrations were 24–42 μeq litre^{-1}. In the 1930s, heavy industrial sites inland showed annual rainwater Cl^- as high as 500 μeq litre^{-1} (BISRA, 1938).

There has been no definitive calculation of chloride emissions in the UK but tentative estimates of the major sources can be made. An average Cl content of 0·25% is appropriate for coal in the UK

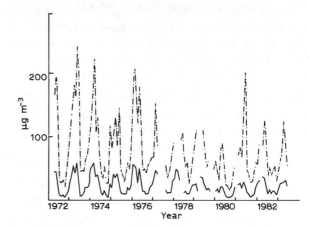

Fig. 4.8. Ozone concentrations in Central London. Monthly average concentrations (——) and 99th percentile of hourly average concentrations in each month (–·–·–).

and assuming 95% of the chlorine is emitted to atmosphere this implies an HCl emission of 240 kt for 1983 from UK coal consumption. Emissions from oil fuels are very low by comparison and an estimate of emissions from this source suggests a figure of 1.5 kt year^{-1} for 1983. The incineration of refuse can also be a source of HCl and estimates of emissions based on WSL measurements at several municipal incinerators suggest a national emission of HCl from this source of 5–7.5 kt year^{-1}. Another source of HCl is likely to be that from the manufacture of chlorine and HCl. Very tentative preliminary estimates of emissions from this source suggests an annual emission of 15 kt year^{-1}. A more detailed assessment of HCl emissions from this source would be necessary before firmer conclusions could be drawn but nonetheless it seems likely that coal combustion represents the major source of national emissions of HCl.

De-icing salts are an important source of chloride with about 2 million tonnes of rock salt estimated to be applied to UK roads each year.

4.3 POLLUTANT DEPOSITION ONTO SURFACES

The two most important mechanisms for transferring pollutants to material surfaces are dry and wet deposition. The complex patterns of air flow and its turbulent structure close to buildings present unique conditions for the deposition of pollutants in urban areas. Some parts of buildings are, because of their location and physical characteristics, very 'exposed' so that driving rain may play an important role in the degradation of such structures. In this section a brief overview of these processes is given.

There is more information available on the dry and wet deposition of sulphur species (chiefly SO_2 and SO_4^{2-}) and the UK Review Group on Acid Rain (RGAR) summarised the information available up to 1981 in its first report (Barrett *et al.*, 1983). This work has been updated in that Group's second report (Barrett *et al.*, 1987). Figures 4.9 and 4.10 are taken from the first report and illustrate the dry and wet deposition of sulphur in the UK. The dry deposition pattern was calculated using a dispersion model and an average deposition velocity for each 20×20 km grid square. The wet deposition map was derived from measurements of sulphate in rainfall, which were not available for many areas in England and Wales. The broad pattern which emerges is that in the areas remote

from major sources and where rainfall is high, wet deposition predominates, but, for areas nearer to major sources in Eastern England where data are available, dry deposition predominates. This is also likely to be the case in major urban areas, although wet deposition data are not plentiful, and the RGAR reported that uncertainties still remain over the dry deposition process in urban areas.

At this stage it is perhaps appropriate to give a brief description of the dry deposition process by which pollutants are transferred from the atmosphere to solid (and liquid) surfaces, a process effected by turbulent transfer. Assuming that transfer close to the surface is governed by diffusivity theory, one may write

$$F = K(z) \frac{\partial \chi}{\partial z} \tag{1}$$

for the SO_2 flux, where $K(z)$, the diffusivity, is a function of height, z, and χ is the SO_2 concentration. It is usual to assume linearity and to define a conventional 'deposition velocity', v_d, such that

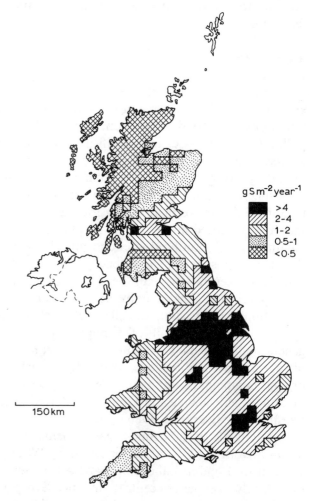

gS m^{-2} year^{-1}

> 4
2–4
1–2
0.5–1
<0.5

150 km

Fig. 4.9.   Dry deposition of sulphur over Great Britain.

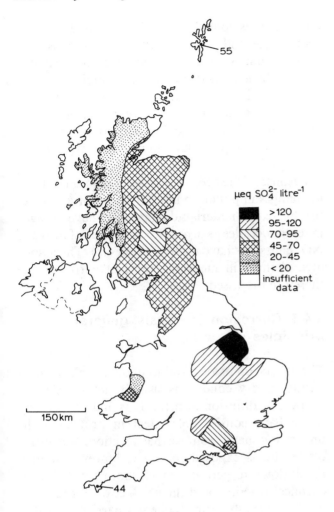

Fig. 4.10. Rainfall weighted annual average non-marine sulphate ion concentrations.

$$F = v_d \chi_1 \qquad (2)$$

where χ_1 is the SO_2 concentration at a reference height, z_1, conventionally 1 m. At some height, a, the SO_2 concentration falls to zero so equations (1) and (2) may be rearranged and integrated to give

$$\frac{1}{v_d} = r_t = \int_a^{z_1} \frac{1\,dz}{K(z)}$$

where the total resistance, r_t, is the reciprocal of the deposition velocity. This equation lends itself to a division of the deposition process into various physical stages, each with its own resistance. The division may conveniently be made into atmospheric and surface processes. The atmospheric processes include an aerodynamic resistance (r_a) which is determined by the wind velocity and aerodynamic roughness of the surface, and a 'bluff-body' resistance (r_b) which is a correction to take account of differences in the transfer of momentum

and gases to rough surfaces (Chamberlain *et al.*, 1984).

In the RGAR report, a value of $17\,\text{mm s}^{-1}$ was suggested as a typical urban deposition velocity for SO_2. Using an interpolation formula for the bluff-body correction (Chamberlain *et al.*, 1984), it appears that not only may the deposition velocity be smaller than the RGAR value but it is also a function of wind speed. Thus for geostrophic wind speeds of 4, 8 and $12\,\text{m s}^{-1}$ the deposition velocity over an urban area of aerodynamic roughness length 1 m would be 2·4, 4·0 and $5·4\,\text{mm s}^{-1}$, respectively.

The value used by the RGAR is the average for urban areas, including all building surfaces and vegetation (which represents a major component of the surface in the residential areas of cities), whereas the values obtained from the bluff-body formula in the previous paragraph may be more appropriate for 'buildings' than for the entire urban area. However, both of these approaches provide only approximate values and the difference may be seen as a reflection of the large uncertainty which exists. More field and laboratory studies are required to resolve these problems.

The significance of the above range of values for the gas-phase resistance depends upon the value of the surface resistance and this is known to depend upon the physical state of the surface, in particular on the humidity and the pH. Liss (1971) found that for SO_2 deposition to a water surface the resistance on the liquid side of the interface was negligible compared to the gas-phase resistance for pH 4. This would presumably also be the case for a film of rainwater on building stone. Spedding (1969) measured the uptake of SO_2 to a nominally dry oolitic limestone surface and found that at a relative humidity of 80% the surface resistance was $0·095\,\text{s mm}^{-1}$ (no separate measurement was made of the aerodynamic resistance of the chamber) while at an RH of 12% this increased by $0·7\,\text{s mm}^{-1}$. Other studies have been made of SO_2 deposition to calcareous soils. Payrissat & Beilke (1975) studied the uptake of SO_2 to a range of soils at various pHs and found that for a soil of pH 7·6 the surface resistance was $0·039\,\text{s mm}^{-1}$ at 44% RH and $0·024\,\text{s mm}^{-1}$ at 88% RH. Similarly, in a study of a chalky field, Garland (1977) found no appreciable surface resistance (i.e. $<0·05\,\text{s mm}^{-1}$) even when the surface appeared to be dry. These measurements would thus appear to be in agreement at high humidities but the resistance of a dry surface seems to be problematic. It is possible that soil and stone have different properties when dry, with the more fragmented structure of dry soil allowing SO_2 to penetrate further to find suitable receptor sites. Taking these values together with the results of the calculations of the gas phase resistance, we may say that in very dry conditions the deposition velocity to limestone is dominated by the surface resistance and may fall to $1\,mm\,s^{-1}$. In moderately humid conditions, the surface resistance will be small ($< 0.05\,s\,mm^{-1}$) but may be significant in high winds or exposed locations, where deposition velocities may exceed $5\,mm\,s^{-1}$. For rain-washed surfaces, only the aerodynamic resistance should be significant.

The occurrence of rain and high winds are not independent, with the frequency of rain in calm conditions being typically a third of that in moderate to strong winds. This, however, should only have a small effect on the overall deposition since even in high winds there is only rain on about 12% of hours. In light winds there is little rain and the aerodynamic resistance is in any case high.

It thus appears that aerodynamic resistance is the principal factor governing deposition of SO_2 in a city, with the gas-phase resistance being typically in the range 0.2–$0.5\,s\,mm^{-1}$. A precise calculation of gas-phase resistance depends on an estimate of the bluff-body correction and its value in a city relies on extrapolation from measurements with roughness Reynolds numbers an order of magnitude smaller. If aerodynamic resistance is the principal limiting factor, it follows that surfaces exposed to the wind will suffer higher deposition.

Although it is well-established that a humid limestone surface (RH 90%) will have a low surface resistance ($0.05\,s\,mm^{-1}$), the behaviour of the surface at lower humidities has not been so well studied. This will be significant for estimating the total deposition in the summer months.

In summary, there are significant uncertainties in the understanding and specification of the turbulent transfer of pollutants to urban surfaces which severely limit the accuracy with which calculations of deposition to building and material surfaces can be made. Moreover, there are further uncertainties in the nature of the variation of deposition velocities with wind speed and wetness of the surface.

4.4 POLLUTANT EFFECTS ON METALS AND COATINGS

This section summarises the available information on the deleterious effects of atmospheric pollutants on the metals commonly used in building and civil engineering and on the surface coatings used to protect them. Since a variety of materials are in use, each with its own degree of susceptibility to corrosion and its own response to different corrosive environments, a general discussion of degradation mechanisms is followed by sections on iron and steel, zinc, copper, aluminium and on protective coatings.

Because of the need to assess the economic effects of changes in pollution levels, it is important to see to what extent materials damage rates have changed in recent decades as urban SO_2 levels have fallen and NO_x emissions have risen (Section 4.2). The associated changes in measured rates of corrosion are therefore discussed.

4.4.1 Corrosion of metals: general principles

The layers of oxide produced on metals by oxidation in dry conditions usually act as effective barriers to diffusion of metal ions and oxygen and hence have considerable effect in protecting the underlying metal from further reaction. When this oxide film is exposed to water, however, it may break down to permit a much more rapid electrochemical reaction, and the product of this reaction does not usually form a coherent barrier to further attack (Evans, 1960; Rozenfeld, 1972). Rates of corrosion are comparatively low even in wet conditions unless solutions of strong electrolytes are present to cause widespread dissolution of the protective oxide film. Exposure to rain, which frequently contains such contaminants as sulphuric acid and sea salt, is not essential for corrosion, since condensed moisture may dissolve corrosive contaminants such as sulphur dioxide, hydrochloric acid and ammonium salts. This is sometimes a problem in factories and warehouses, and it may also affect building components. Also, since many of these species are hygroscopic or form hygroscopic corrosion products, liquid films may be formed in atmospheres well short of saturation with water vapour. It is usually considered that significant corrosion takes place only above some critical value of relative humidity, which varies with the metal and the corrosive substances present. Metal specimens placed inside screens which shield them from direct contact with rain, but allow access of air and dust, usually corrode much more slowly than specimens openly exposed at the same site. At very damp sites with considerable pollution with particulates, shielded specimens may, however, corrode faster because contaminants are not washed away by rain. Rates of corrosion are often still lower inside buildings, where heating reduces the relative humidity and largely prevents formation of dew. Oxygen is also essential to the corrosion process except in very strongly acid solutions. Rates of corrosion therefore depend on the time during which the metal is wetted (or during which the critical relative humidity is exceeded), on the nature and concentration of the pollutants present, and on the supply of oxygen. The corrosion of building components can therefore be expected to depend on details of exposure and design. The effects of exposure to wind, sun and rain, the presence of moisture traps and proximity to local sources of pollutants will be important, for instance. It has been suggested that for a given 'time of wetness' (measured by electrochemical devices, or estimated from average relative humidity) the actual extent of corrosion may vary with the number of wet/dry cycles, and that there may be important effects due to seasonal variations in rainfall or pollutant levels.

It is therefore not easy to design exposure trials to establish representative rates of corrosion in a particular district, or to use such data to predict the life of a building component. It is more difficult to encapsulate measured corrosion rates in 'damage functions' for predicting the rate to be expected in a given environment.

The effects of atmospheric exposure on the corrosion of metals can be investigated in the field or in environmental chambers in the laboratory. Field measurements are complicated by the fact that important variables may not have been measured, and those that have been measured may not have been adequately controlled or may not be adequately represented by a single average value. There may

also be unexpected interactions between factors. Laboratory environmental chambers inevitably fail to simulate all the features of the natural environment, particularly the natural cycles of temperature and humidity. They have the advantage of being able to isolate the effect of a particular factor and to provide reasonable first approximations for the functional dependence of corrosion rates on the various parameters. They may also be used to obtain test data in a shorter time than would be possible in field trials, even though comparisons between rates in accelerated tests and in natural exposure are often hazardous.

Figure 4.11 shows the results of several experimental determinations of the rate of corrosion of zinc in air with various concentrations of sulphur dioxide. Not only is there a large scatter in the rates, but the functional dependence on sulphur dioxide concentration also shows large variations from one series of experiments to another. These variations are due to uncontrolled variations between the different experiments. Figure 4.11 includes the results obtained with galvanised steel placed in different parts of a field which were fumigated with different concentrations of sulphur dioxide, but in other respects were subjected to the same conditions.

It is probably necessary to combine information from field and laboratory studies in order to arrive at satisfactory damage functions.

Little work has been carried out on the derivation of damage functions appropriate to the climatic conditions in the UK, apart from Shaw's (1978) survey on zinc (see Section 4.4.3.2). Table 4.1 collates measurements on the rates of corrosion of ingot iron, a copper-bearing steel and zinc, obtained by the British Iron and Steel Research Association (BISRA) and subsequently by the British Steel Corporation (BSC), from the 1930s onwards. It illus-

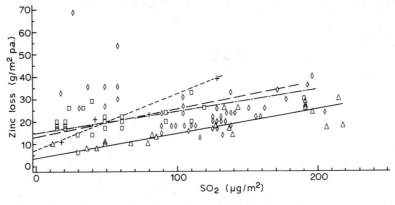

Fig. 4.11. Corrosion of zinc at different SO_2 levels including recent data from zinc cans exposed in a field where SO_2 was varied, but other factors remained constant (Bawden & Ferguson, 1987). Key: ◇ --- ◇, Haynie & Upham (1970); □ — — □, Knotkova-Cermakova *et al.* (1974); △——△, Guttman (1968); + - - - +, Bawden & Ferguson (1987).

Table 4.1 Corrosion rates measured by British Iron and Steel Research Association from 1930s onwards

Site	Rates of corrosion (μm year^{-1})				Rainfall (mm year^{-1})	Rain analysis			Air SO$_2$	
	Ingot iron (1 year)	Ingot iron (5 years)	0·3% Cu steel (1 year)	Zn		Cl$^-$ (μeq litre^{-1})	SO$_4^{2-}$ (μeq litre^{-1})	pH	PbO$_2$ candle mg SO$_3$ per 100 cm^2 day	Equivalent SO$_2$ (μg m^{-3})
Brixham[a]	53	56	36	1.8	148	151	57.5	6.5		
Llanwrtyd Wells 1932–37	63·5		43	3	62	595	172	6.9		
Calshot 1933–37	79	114		3.6						
Motherwell[a]	97	61		4.3						
Woolwich 1931–37	102	69	76	3.8	43	244	1014	4.9		
Birmingham Suburb[a]	104			3.6						
Sheffield University[a]	114	86		4.6						
Attercliffe 1934–37	137	119	97	16	51	502	431	4.3		
Brown–Firth Lab[a]	145		97	11						
Derby (Old site)[a]	130			7.6						
(New site)[a]	173		109	7.1						
Frodingham[a]	163			10						
Godalming[b]	48		42	1					0·23	27
E Greenwich[b]			101	5.8					2·2	255
Sheffield (Hunshelf Bank)[b]			107	13					3·58	414
(Sewage Works)[b]			108	11					3·66	424
Dove Holes Tunnel[a]	81	64		81						
Rye, 1979–80			59·3						0·46	53
Stratford, 1979–80			62·5						0·99	115
Stratford, 1983–84			37·6						0·353	41
Brixham, 1979–80			33·8						0·40	46
Mickle Trafford, 1979–80			47·4						0·47	54
Khartoum 1930s	1·2			0.5						
Abisko 1930s	4·6			0.5						
Nkopku 1930s	5		5	0.5						

[a] Hudson & Stanners (1953) 'Up to 20' one-year trials pre-1953.
[b] Hudson & Stanners (1951).

trates a number of points made in the discussion in that rates of corrosion are extremely small in arid conditions (Khartoum) and also in a sub-polar climate where there is little pollution and water remains frozen for most of the year (Abisko, in Northern Sweden). Rates are considerably higher at rural sites in temperate climates, and much higher again in industrial areas with heavy pollution with SO_2 and Cl^-. It is clear that damage functions must reflect these effects, the total corrosion being expressed as the sum of a series of corrosion events, each with average rate appropriate to the conditions of temperature, humidity and pollutant level.

A large number of expressions relating the rate of corrosion of various metals to climatic and pollution data have been obtained by curve-fitting results from corrosion experiments. Each such function is inevitably constructed using the quantities measured in that particular trial, and the data set seldom includes all the parameters that might be relevant. It may happen that a certain parameter remained constant throughout, or was highly correlated with some other quantity, so that the published function is not applicable at other sites. Damage functions tend to involve a small number of variables selected from a large number of possibilities, and they differ considerably both in the variables considered and in mathematical form. The use of these functions for predicting rates of corrosion at sites other than the ones at which the data were obtained therefore involves assumptions which are often questionable. Among these are the spatial variability of pollutant levels and the averaging of time-dependent variables. Most site investigations quote average rates obtained by measuring loss of metal over a fairly long period of time.

The use of damage functions to estimate economic cost requires the use of atmospheric data averaged over large areas and obtained from computer modelling, so that only the simplest functions can be considered. Even then it is not clear that time averages can be employed, and the stock at risk and the pollutants are probably concentrated in a small part of each cell, so that the spatial averaging underestimates the true extent of the damage. The accuracy of this type of calculation is therefore likely to be low.

The damage functions available for building materials have been assembled in various comprehensive compilations, the major of which include Nriagu (1978), Umweltbundesamt (1980), UNECE (1984) and Jorg *et al.* (1985).

These reviews show that fifteen or so different functions are available for steel, and a similar number for zinc or galvanised steel. A smaller number exist for stone and paint. Jorg *et al.* (1985) assembled 998 references on materials damage due to air pollution and besides listing the damage functions therein contained, provide a cross reference system for finding the relevant references for each material–pollutant combination.

4.4.2 Pollutant deposition onto metal surfaces

4.4.2.1 SO₂ deposition

Considering the deposition of sulphur species in urban areas, although measurements are less plentiful than in rural locations, in urban areas dry deposition of sulphur dioxide is probably the major component of sulphur deposition. When annual average concentrations were $400 \, \mu g \, m^{-3}$ this mechanism (assuming a deposition velocity of $8 \, mm \, s^{-1}$) could give rise to a deposition of $50 \, gS \, m^{-2} \, year^{-1}$. Early measurements of wet deposited sulphate using bulk deposit gauges exist for London and other areas of the UK from 1920. These must be treated with some caution not solely due to the fact that bulk deposit gauges are open to contamination from extraneous sources (e.g. birds, etc.) and include some dry deposition but also because of the likely low level of quality assurance and rigorous sampling and analysis protocols at the time. However, the data suggest annual wet deposition rates of $5–10 \, gS \, m^{-2} \, year^{-1}$.

SO_2 may be deposited onto metal surfaces by either 'wet' or 'dry' processes, i.e. in rain or by absorption on the surfaces of building materials, etc. Most wet deposition of sulphur dioxide takes place at considerable distances from the source, and short-range deposition involves mainly dry processes, which account for about two-thirds of total deposition in the UK. The proportion of dry deposition in towns is much higher than the national average. Since metallic objects are concentrated in towns, it seems clear that most damage to metals involves dry deposition of pollutant emitted by local sources and discharged through comparatively low chimneys. Sulphur dioxide makes the largest contribution to this damage.

4.4.2.2 NOₓ deposition

Considerably less is known about the deposition of nitrogen oxides. There is little information on their effect on the corrosion of metals, but it seems likely that nitrogen dioxide may stimulate the corrosive effect of sulphur dioxide even if it does not produce very marked effects by itself (see Section 4.4.3). Except at isolated sites (e.g. eight years at a site in the North-east Midlands) few measurements of NO_x levels have been made in the UK until comparatively recently, and practically no exposure trials have provided data for NO_x. It is therefore not possible to comment on the extent to which NO_x contributes to the overall degradation of building materials, and much more work is needed in this direction.

4.4.2.3 Chloride deposition

All rain in the UK contains sufficient chloride derived from sea-salt to cause appreciable corrosion of iron, copper and zinc and this tends to impose a lower limit on rates of corrosion irrespective of pollutant levels. Rates of corrosion are significantly higher close to the coast than at rural inland sites, although marked effects due to chloride have been observed at especially windswept sites 40 km inland. Although damage functions have been published quantifying the effects of chloride there is considerable disagreement in the literature on the magnitude of the effect. Typical figures for the chloride content of rainwater are included in Table 4.1.

Hydrogen chloride is produced in some industrial processes, particularly by the combustion of high-chlorine coal. The air in industrial towns was, in the past, seriously contaminated with hydrochloric acid, and although industrial emissions have declined, there is probably still a significant contribution in northern towns from local coal burning. In all towns, a major source of chloride is road salting. The most important source of hydrochloric acid is electricity generation, which consumes about two-thirds of total coal production. Since hydrochloric acid is very soluble in water, it is deposited comparatively close to the source when rain falls through a plume. Power stations are, however, for the most part no longer sited close to, or to windward of, large towns, so that the impact of the hydrochloric acid emitted by power stations on urban metalwork is probably small.

4.4.2.4 Deposition of particulates

Besides ammonium salts and salts derived from sea water, rain in the UK contains insoluble matter derived from soil and rock residues and from combustion processes, including soot, acid smut and fly ash. Such material may therefore be deposited in the

form of aerosol or in solution or suspension in raindrops or dew. Vernon (1935) showed many years ago that atmospheric corrosion could be largely prevented even in contaminated atmospheres merely by excluding particulate matter. Many studies have shown that deliberate contamination with salt particles can stimulate corrosion and both salt particles and carbon can enhance the effect of gaseous pollutants. Ammonium salts, which form an important part of surface contamination, can be very effective in initiating corrosion, probably because they produce solutions of low surface tension which are much more acid than the corresponding sodium salts. The effects of chlorides in enhancing the effect of sulphur dioxide or sulphates, which have often been reported, are probably due to the increased time of surface wetness due to deliquescence.

Several workers have indicated the additive effects on corrosivity of sulphur dioxide/sulphates and chlorides in mixed environments. Thus it is to be expected that rates of corrosion will be correlated with contents of sulphur dioxide and moisture in city air. (Urban deposition is approximately $10\,g\,SO_2\,m^{-2}\,year^{-1}$ and $1\,g\,Cl^-\,m^{-2}\,year^{-1}$.) Other authors noted that the role of sulphur dioxide appeared prominent even in mixed industrial and marine conditions (Tulka & Schattauerova, 1982; Pourbaix, 1982; Johnson, 1982). Synergism has been found between sulphates, sulphur dioxide and sodium chloride with stainless steels, copper and aluminium in natural environments, and in laboratory work using (solid) contaminants of carbon, sodium chloride, ammonium sulphate, ferrous sulphate, and (gaseous) sulphur dioxide (Kamaya et al., 1981; Feliu et al., 1984). There were additive effects between chlorides and sulphur dioxide in the corrosion of aluminium, copper, iron and zinc (Gonzalez & Bustidas, 1982). In contrast to the above studies, other researchers have emphasised the importance of the chloride contribution to corrosion in a mixed pollutant environment. Such works have shown correlation coefficients in excess of 0·77 (over 0·9 in winter), between corrosion rates and atmospheric chloride concentration, and the significant corrosion-promoting effects of chloride species even when present in low concentrations (Knotkova-Cermakova & Marek, 1976; Yasukawa et al., 1980; Himi, 1981; Corvo-Perez, 1984). The aggressivity of the chloride-containing particle or aerosol is very important and expected, and it is worth noting two marine-based exposure programmes (Gullman & Swartling, 1983; Cordner et al.,

1984). With chloride deposition in rain about $22\,g\,m^{-2}\,year^{-1}$ and dry deposition approximately $6{\cdot}5\,g\,m^{-2}\,year^{-1}$, it was noted that prevailing wind direction and speed could be more important than actual distance inland in determining deposition rates. This particularly related to dry deposition, and it was further observed that such deposition was an important factor in the corrosion of sheltered surfaces. Data for aluminium, copper, steel and zinc showed that, after one year's exposure, the corrosion rate was a linear function of the chloride content of the atmosphere.

The effect of otherwise inert particles in providing a favourable environment for corrosion by absorbing moisture or sulphur dioxide, or by forming a crevice where oxygen concentration is low, or by supplying catalytic species, is also of considerable importance. Fly-ash, smut and smoke particles have been shown to produce a number of these effects, and may also act as a source of corrosive salts (Walton *et al.*, 1982*a*, *b*).

4.4.3 Survey of corrosion rates

4.4.3.1 Ferrous alloys
The first systematic field tests of metallic materials in the UK were started by BISRA in the 1930s and continued until the late 1950s. These tests clearly indicated the increase in corrosion rate with increasing ambient sulphur dioxide concentration for sites with comparable total times of wetness. Data for a plain carbon steel and a $0{\cdot}3\%$ Cu low alloy steel are shown in Table 4.1 (BISRA, 1938; Hudson & Stanners, 1953).

Plain carbon steel has been used in many other national and international field studies for comparing the relative corrosiveness of different environments. Empirical expressions have been published relating metal loss by corrosion to environmental parameters. These expressions invariably involve the sum or the product of terms containing ambient SO_2 concentration and some quantity related to atmospheric humidity. The functions are often non-linear. The corrosion product formed in the presence of SO_2 contains ferrous sulphate, which is stored in the scale, so that 1 molecule of SO_2 can convert many Fe atoms to rust (Schickorr, 1963; Barton & Bartonova, 1969).

The studies so far described have not related corrosion rates to the pH (or composition) of the falling rain. In the analyses most recently reported, Haagenrud *et al.* (1985) found it unnecessary to invoke rain pH in addition to SO_2 concentration

and time of wetness to explain field data in Norway. However, Lipfert *et al.* (1985) have suggested a non-linear damage function that includes a significant pH term for use in the US National Acid Precipitation Assessment Program. Future field and environmental chamber studies should be designed to resolve this question.

Johansson (1985) has investigated the corrosion of carbon steel in environmental chambers where the air was dosed with SO_2, with and without NO_2. Nitrogen dioxide enhanced the SO_2 pick-up and steel corrosion rate, but only at low (50%) relative humidity, apparently by producing a hygroscopic surface, i.e. by reducing the critical relative humidity. This may not be significant if hygroscopic particles are present. Damage functions have not been reported for highly alloyed stainless steels in polluted environments. Even when sulphur dioxide concentrations are high, only slight discolouration or shallow pitting has been observed (Johnson, 1982).

Plain carbon steels are not generally used in structures exposed to the atmosphere in unprotected form. Corrosion arising from failure of the protection (e.g. paint or galvanising) tends to be localised. Structural damage may result not only from loss of cross-section by corrosion but by the mechanical action of a growing corrosion product which occupies a volume greater than that of the original metal (Harris, 1984).

4.4.3.2 Zinc
Steel is commonly protected by coatings of zinc 50–$150\,\mu m$ thick, which corrode much more slowly than the steel with the formation of a layer of insoluble corrosion product. Some benefit is obtained from galvanic protection at breaks in the galvanising, since zinc is anodic with respect to steel. Unlike mild steel, zinc and zinc-coated steel are often exposed to the weather without further protective treatment, although for critical applications, paint schemes are of course applied over galvanising.

Numerous field studies have indicated that sulphur dioxide in the atmosphere can significantly increase the rate of zinc corrosion. In clean rural areas ($SO_2 < 10\,\mu g\,m^{-3}$) it is only $1\,\mu m\,year^{-1}$, but the corrosion rate rises to $> 5\,\mu m\,year^{-1}$ in industrial atmospheres (with $SO_2 > 100\,\mu g\,m^{-3}$)(Shaw, 1978).

A major field study of the corrosion of zinc in the UK was organised by Shaw. A pilot study on CEGB transmission towers was started in 1966 and was expanded to include 3000 zinc-can samples placed at locations around the UK where weather data were also available (Shaw, 1978). The 1969–75 data form the basis of the atmospheric corrosion map published by the Ministry of Agriculture, Fisheries and Food (MAFF, 1982). Saunders (1982) has analysed the original Shaw data to produce a UK zinc damage function.

With some exceptions, field investigations have suggested that zinc corrosion is a linear function of both sulphur dioxide concentration and time of wetness (defined typically as time when RH > 80%). Figure 4.11 shows data from a number of field studies. The trend towards increasing corrosion loss as the ambient sulphur dioxide concentration is raised is clear, but there is considerable scatter arising from different times of wetness and variations in other environmental variables. Figure 4.11 also shows data from a CEGB study in which the SO_2 concentration was varied artificially across a field site, other conditions being the same. A simple dependence on SO_2 concentrations is evident.

Johansson (1985) included zinc in environmental chamber studies of the effects of NO_2 and $NO_2 + SO_2$ on atmospheric corrosion. Though NO_2 alone had no effect, it considerably enhanced the effects of SO_2 when present at concentrations of $560 \mu g\, m^{-3}$ or higher. It is possible that NO_2 acts as an oxidant for SO_2 in an acid-forming reaction of the type

$$SO_2 + NO_2 + H_2O \rightarrow H_2SO_4 + NO$$

and nitric oxide production was observed in the test chamber. Further work is required in order to establish whether NO_2 has a significant effect at the lower concentrations usually encountered. If it does, this may account for some of the scatter in corrosion data from field studies.

Sea salt affects the corrosion of zinc up to 10 km inland and HCl gas from industrial sources may also have a small effect on the rate (Bawden & Ferguson, 1987). Recently it has been suggested that formaldehyde in the atmosphere may have a corrosive effect on zinc (Graedel, 1986).

The direct effect of rain on zinc corrosion is less well understood. Haagenrud *et al.* (1985) in their most recent analysis of field data obtained in Norway found it unnecessary to invoke a separate term with rainfall parameters. However, Saunders (1982), in her analysis of UK data, found that annual rainfall intensity was significant. In a study in a low-SO_2 area of the United States, Spence *et al.*

(1985) also found that corrosion weight losses of galvanised steel panels decreased by a factor of two if the panels were sheltered from the rain. They took this to mean that rain-washing has a direct effect in addition to its influence on time of wetness, and this is consistent with Saunders' analysis. Neither study shows whether rain composition or pH is important.

4.4.3.3 Copper
Copper is often used as a roofing material on public buildings, the familiar green patina arising from atmospheric corrosion. Field studies generally show corrosion rates to be about half those of zinc under comparable conditions, i.e. $0.5 \mu m\, year^{-1}$ in a clean rural area, increasing to $2 \mu m\, year^{-1}$ in a moderately polluted industrial area. These rates are low, and copper roofing has a long life even in industrial atmospheres, especially as it tends to be used in sections much thicker than the zinc coatings on steel.

SO_2 has been implicated as the industrial pollutant responsible for enhanced corrosion of copper in short-term studies (Guttman & Sereda, 1968). However, Lipfert *et al.* (1985) have suggested that SO_2 sensitivity decreases after the first year of exposure and that chloride is a more important pollutant.

The copper 'skin' of the Statue of Liberty was found to be in good condition with only 10% of its 2 mm thickness corroded away in spite of a century of exposure to an urban, marine environment.

4.4.3.4 Aluminium and its alloys
Pure aluminium and its dilute alloys are generally very resistant to atmospheric corrosion. The corrosion rate decreases with time due to the formation of a stable aluminium hydroxide film. In rural areas general corrosion rates are $< 0.1 \mu m\, year^{-1}$, and they increase to only $0.2–1.0 \mu m\, year^{-1}$ in heavily polluted industrial areas. Pitting attack is observed to greater depths under all conditions, but is usually acceptable. Aluminium and its alloys are therefore generally regarded as having good corrosion resistance in atmospheres polluted by sulphur compounds, and derivation of damage functions has been considered unnecessary.

Accelerated linear corrosion of aluminium in galvanic contact with carbon or steel has been observed on CEGB electricity transmission lines, with localised pitting due to carbon particles, and galvanic corrosion under graphite-loaded rubber bushes. In addition a more severe corrosion of aluminium in contact with steel has occurred on transmission lines, with the highest rates near to the coast (Jackson *et al.*, 1987). The principal corrodent in this form of attack is chloride, either from sea salt or from HCl emissions from industrial sources.

4.4.3.5 Paint

Paints provide economical and versatile organic coatings, suitable for both new equipment and maintenance and repair operations. They are used for both protection and decoration on metals and also on wood and other non-metallic materials. The usefulness of a paint film depends on a complex set of properties, including water-exclusion, the adherence, flexibility and physical robustness of the film, the reactions of pigment particles (including fading, chalking, etc. and chemical properties), resistance to moulds and fungi and corrosion-inhibiting properties. The integrity and adherence of paint films is extremely important, and even small breaks or cracks in an otherwise sound film may lead to the formation of pockets of moisture which, by keeping the substrate continually wet, may accelerate failure, not only of the paint but of the substrate itself. The action of paints in protecting metals from corrosion does not depend primarily on excluding oxygen and water from the metal surface: their main effect is to exclude corrosive salts, and priming coats also supply corrosion inhibitors which may give some protection to the metal under gaps or damaged areas in the paint. Thorough surface cleaning is essential as a preparation for painting or repainting, for if corrosive salts (which are present in rust) are painted over, the supply of water and oxygen diffusing through the coating will be sufficient to set up corrosion cells leading to loss of adhesion and blistering. It is desirable to use protective metal layers or phosphate conversion coatings as a preparation for painting (British Standards Institution, 1977, 1982).

Paint films are slowly attacked by photochemical processes, and the filler particles may take part in these reactions. There is little evidence that atmospheric pollutants play a major role in paint damage mechanisms. The most important processes are photo-oxidation by UV radiation and by ozone. However, SO_2 has been reported to impede the drying of certain paints and some pigments, and particularly extenders based on calcium carbonate, may be attacked by SO_2. There is also some evidence that certain paint vehicles may be hydrolysed by highly acidic solutions. Survey evidence suggests that although repainting cycles are shorter in city centres, this is due more to considerations of prestige than to any marked effect on paint life, and it is by no means clear that painting schedules correspond to the economic life of the paint system. It is probably true that once corrosion has begun under a paint film, the remaining useful life is shorter in a more corrosive atmosphere, but paint films are normally renewed when a few percent of the surface is rusted.

Damage functions (Fink *et al.*, 1971) imply that paint life is shortened by one year by an increase in SO_2 content of $62.5\,\mu g\,m^{-3}$ or $32\,\mu g\,m^{-3}$ on plain steel and galvanised steel, respectively, which confirms that the effect of present levels of pollution on the economic life of paint films is not very great.

4.4.3.6 Polymer coatings

A range of polymer coatings are used on steel sheet, using roll-bonding, plastisol coating, and stoved dispersions. They are often applied to galvanised steel for use in cladding. Similar considerations apply as for paints. There is little evidence that the life of these coatings is seriously affected by atmospheric pollutants although, again, failure may be accelerated once the coating has been penetrated. Some problems are discussed in the review by Hudson (1986).

4.4.3.7 Reinforced concrete

Steel reinforcing bars in concrete are effectively protected by the highly alkaline cement paste unless the latter contains excessive quantities of corrosive salts from inadequately washed aggregate or setting accelerators. The tolerance for contaminants is limited, however, and corrosion of the reinforcing rods is always possible in cracked and porous regions of the concrete by entry of sea-spray or road de-icing salts. Because the corrosion product is much more bulky than the steel, corrosion, once started, causes cracking of the concrete with exposure of fresh steel surface.

Carbonation of the cement paste by atmospheric CO_2 over a period of years causes the pH to fall so that the reinforcement is no longer adequately protected. For this reason BS 8110 (1985) specifies minimum grades of concrete, and minimum nominal cover for reinforcement increasing from 20 mm in 'mild' environments to 60 mm in 'extreme' conditions.

Since the SO_2 content of the atmosphere is very

much lower than the CO_2 content (about 20 ppb (10^{12}) compared with 0·03%), the effect of SO_2 in decreasing the pH of cement paste is scarcely significant, and the most important factors in corrosion protection are the integrity of the concrete and the exclusion of sodium chloride. Once the reinforcement has been exposed, it is possible that the remaining life may be reduced in heavily polluted atmospheres, but in most cases it is likely that the situation will still be dominated by salts.

Skoulikidis (1982) found that the thickness of the rust formed on reinforcements in concrete dismantled after periods of time between 15 and 140 years fell on a single parabolic curve even though they came from sites with differing levels of pollution over a period in which pollution was changing considerably with time. This tends to confirm that atmospheric pollution may not be a major factor determining the failure of reinforcements.

4.4.3.8 Metallic ties in masonry and brickwork ·
Many metallic components are used to give stability to masonry and brickwork. These components need to be protected against corrosion since their environment often ensures that once water is allowed to enter drying takes place very slowly, if at all.

Galvanised wall-ties may rust because of faulty detailing which allows water to run along the metal and keeps it permanently wet. Ties for masonry should be designed and mounted to prevent ingress of moisture; historically, they have often been set in lead.

Corrosion arising from the use of insufficiently resistant or inadequately protected materials can lead to the bursting of the masonry. Perhaps the most celebrated example of this effect is the expansive damage caused by the rusting of steel reinforcements installed in the buildings on the Acropolis earlier this century. These reinforcements are being replaced in turn with components made from a titanium alloy.

The most important factors determining life are likely to be the quality of the detailing and the integrity of the structure, although as with reinforcements in concrete, it is possible that once the structure begins to admit water subsequent corrosion will be more rapid in polluted atmospheres.

4.4.4 Historical trends in damage to metallic materials

Although it was well known early in the 20th century that rates of atmospheric corrosion increased in the order rural < marine ≤ industrial, and trials of competing products were organised on this basis, no measurements of rates of corrosion appear to have been made at sites with systematic measurement of pollution before the BISRA trials in the 1930s which included rainwater analysis (Table 4.1). The BISRA (later BSC) investigations continued until the late 1950s, and in a small number of cases to the present day. They show a considerable reduction in rates of corrosion of steel since the 1950s. Some measurements of air SO_2 levels made by the lead dioxide candle method in the early 1950s are quoted in Table 4.1. Comparison with measurements made by H_2O_2 bubblers is extremely imprecise, since the results are strongly affected by wind velocity and other factors. An approximate conversion based on a deposition velocity of 8 mm s^{-1} is given in Table 4.1, and the results are again broadly consistent with the overall picture.

As an example of the changes experienced at a particular urban site, BSC measured SO_2 using PbO_2 candles at Stratford in East London equivalent to 46·5 μg m^{-3} (summer average) and 188 μg m^{-3} (winter average) in 1976–79 (overall average 122 μg m^{-3}). These values show large falls compared with industrial districts in London in the early 1950s, but Stratford still stands out as a contaminated area compared with the London average of 65 μg m^{-3}. By 1984, however, the Stratford annual average had fallen to the equivalent of 41 μg m^{-3}. The rate of corrosion of a 0·3% Cu steel fell by 40% between 1979–80 and 1983–84, in line with the reduction in SO_2 level, and the 1984 rate was among the lowest ever recorded in the UK for this type of steel. Similar results were obtained at Sheffield and Rotherham. It is therefore clear that rates of atmospheric corrosion of steel have declined considerably since the 1930s, and that there has been a marked improvement in some districts since 1979–80 in consequence of large reductions in local SO_2 levels. It has been suggested that these changes, which reflect not merely a decline in industrial activity but also changes in fuel usage and the pattern of emissions, may have increased the acidity of rain at rural sites whilst values in towns may have fallen somewhat. The evidence on this point is inconclusive and no published evidence has been found for any increase in rates of corrosion in rural areas.

4.5 POLLUTANT EFFECTS ON NON-METALLIC BUILDING MATERIALS

4.5.1 Pollutant deposition and chemical weathering processes

Solution weathering of limestone by wet deposition is critically influenced by the presence of CO_2, SO_2 and NO_x. Where pH is 5·6 or greater, carbonic acid may be important in weathering carbonates. Where pH falls below 5·6, as is common in polluted atmospheres, acidic aerosols in wet deposition can dissolve carbonates to produce by-products such as gypsum (hydrated calcium sulphate).

But, while it is clear that CO_2, SO_2 and NO_x may all contribute towards solution weathering, we find that the relative roles of these gases are unknown, that CO_2 is rarely monitored anyway, and that the influence of NO_x has yet to be satisfactorily established. SO_2 is undoubtedly significant because of the sulphates it leaves behind.

Sulphation involves the dry deposition reaction between SO_2 gas and the limestone itself, in the presence of high relative humidity and a catalyst (e.g. a metal oxide such as Fe_2O_3, or even NO_2). This also produces gypsum. Of crucial importance here in the context of the 'acid rain' debate is the extent of sulphation that occurs by dry deposition of SO_2, compared with that due to wet deposition.

The more obvious non-metallic building materials likely to be affected by acid deposition are calcareous stones such as limestones, magnesian limestones and calcareous sandstones. There is also a possibility that glass, organic materials (some of which are 'filled' with inorganic oxides), brick, cement, mortars and concretes also may be affected. Important features to be considered in each case are the durability of the materials in the absence of pollutants, which factors are important in 'natural' decay or weathering processes and how these processes are affected by the presence of pollutants or acid deposition. Apart from the chemical reactivity of stones with acidic pollutants, there are other parameters to be taken into account in assessing potential damage such as the porosity of masonry materials, meteorological variables such as windspeed, relative humidity, levels and frequency of precipitation (and consequently how often the surface of a material is significantly wetted). There will also be different mechanisms and sequences of events by which pollutants in different physical forms will attack building materials.

In the case of a limestone, for example, the possibilities include wet or dry absorption of pollutant gas followed by chemical reaction and dissolution of the products in rainwater and direct dissolution of the stone in acidic rainwater (both of these operating on a relatively short time-scale). A much longer term but equally important process is that in which reaction products more soluble in water than the original material (for example, sulphates) move into the body of the stone where they can crystallise and form hydrates in expansive reaction, leading to physical disruption of the stone surface and hence to decay.

4.5.2 Physical weathering processes

Physical weathering processes affect porous materials and take many forms. Brick and some building stones are affected by frost, usually when the moisture content of the masonry is high for prolonged periods, and by transmission of soluble salts from the earth or from marine or other sources. The detailed mechanisms of damage by frost or salt crystallisation are not completely understood. There are some similarities and some differences. The expansion of water on freezing explains the bursting of uninsulated pipes, and this mechanism will also have an effect in frost damage in brick and stone. The simple expansive reaction of ice upon freezing is not the entire story of how stone decays. It is known, for example, that non-reactive materials absorbed into stone which do not undergo expansion on freezing (for example, nitrobenzene) can still have an adverse effect upon freezing. Frost damage and salt damage to stone depend, among other things, on the pore size distribution of the material, its chemical composition, and its mechanical properties. It is known that limestones which have a high percentage of very small pores in them are less durable than those which do not.

The physical manifestations of frost damage to a stone are somewhat different from damage by other mechanisms such as salt damage. Large sections of the surface of frost-susceptible stones may split away after cracking from the bulk of those parts of stone buildings likely to be exposed and to become very wet (for example, balustrades, parapets).

Salt weathering, the physical disintegration of stone by pressures arising from crystal growth of salts from solution and related processes, is important in many buildings, especially where the salt has been produced in the stone as a result of solution and sulphation, or has been deposited as aerosols (as with sea-salt deposition at the coast).

Although the rudiments of these and other processes are understood, much is unknown. The relative importance of the processes or how their relationships change over space and time is not clear, for instance. Little is known about the most significant relationships of all between the intensity of atmospheric pollution, the rates of pollution take-up in stone and other porous materials, and the rates of stone decay (i.e. 'dose-response' relationships).

Studies at St Paul's Cathedral illustrate the consequences of weathering processes on stone (Fig. 4.12). Here, several weathering zones are shown, each of which has distinctive weathering conditions and is associated with distinctive weathering features. For example, there are solution features in run-off zones, such as pitting; solution weathering in drip zones; deposition of gypsum and calcium carbonate in flow zones; cumuliform growths of gypsum and soot in sheltered zones of dry deposition; and case-hardening, flaking and blistering associated often with salt efflorescence. Damage can be especially severe where the ratio of surface area to stone volume is high, as in carved stone. Given such a variety of features on one building in one stone, it is understandably difficult to generalise precisely about the rates at which a building

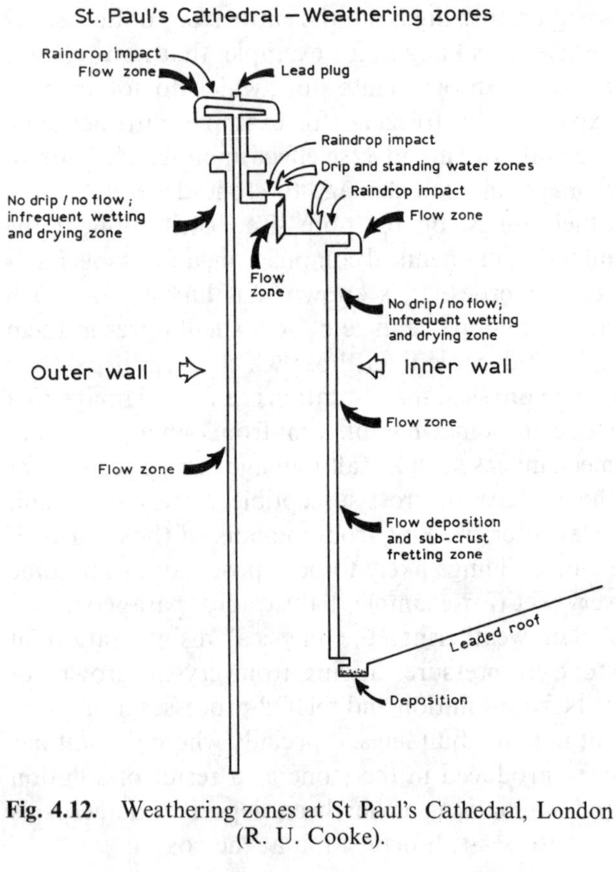

Fig. 4.12. Weathering zones at St Paul's Cathedral, London (R. U. Cooke).

weathers, not least because present-day behaviour is often preconditioned to an uncertain extent by earlier weathering under different conditions. This is the 'memory effect' of stone and is fundamental to all discussions of accelerated pollution and weathering on historic buildings.

4.5.3 Microclimate effects on buildings and structures

The weathering system as it relates to buildings comprises three main groups of variables—those relating to the environment, the building materials and the nature of the building itself. Each group is extremely complex and includes major difficulties for research.

The atmospheric environment involving gaseous, liquid and particulate components, and both wet and dry deposition, is far from being fully monitored. More importantly, the microclimate of weathering at the crucial interface between the atmosphere and the building surface is never adequately monitored and little is known about the ways this environment has changed. Recent studies that have attempted to simulate the stone-atmosphere system in the laboratory are helpful, but the translation of laboratory results to field conditions can be perilous.

Building characteristics are equally fundamental in influencing the nature of weathering. Certain parts of buildings are more likely than others to become saturated with water and therefore to be more susceptible to weathering, for example, cornices, coping stones, plinths and steps. Leary (1983) recognised four exposure zones in buildings (Fig. 4.13), ranging from Zone 1 which is most vulnerable to weathering and requires the use of the most durable stone, to Zone 4, the least vulnerable (areas of plain walling). One must avoid the inappropriate juxtaposition of incompatible materials, and faulty craftsmanship such as 'face bedding' in which bedding laminae in a stone are placed parallel to the building surface and may eventually exfoliate.

4.5.4 Survey of damage rates

If rates of weathering are influenced by atmospheric pollution, measured rates should vary, both spatially and with time. Recent work has been directed towards exploring these initial hypotheses with respect to limestones.

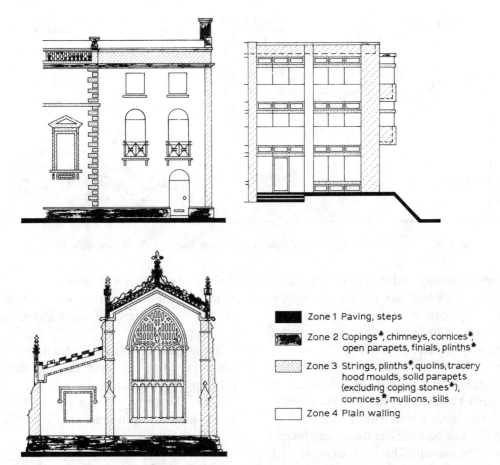

Zone 1 Paving, steps

Zone 2 Copings*, chimneys, cornices*, open parapets, finials, plinths*

Zone 3 Strings, plinths*, quoins, tracery hood moulds, solid parapets (excluding coping stones*), cornices*, mullions, sills

Zone 4 Plain walling

Fig. 4.13. Exposure zones of buildings relating to susceptibility of stone weathering (Leary, 1983). (*A stone normally suitable for Zone 3 could be used for copings and cornices in Zone 2 if it were protected by lead. Similarly, a plinth in Zone 2 could be considered as Zone 3 if there were protection against rising damp.)

4.5.4.1 Spatial variability

Limestones weather naturally and rates of limestone weathering have been measured in supposedly pollution-free areas of the British Isles for periods extending back to the last glaciation (Table 4.2). These data reveal considerable variability (from 3 to 88 μm year^{-1}) that no doubt reflects variations of climate, rock lithology and accuracy of measurement. They provide an indication of the magnitude of 'natural' rates, and their variability for the same rock type. The data provide a warning that there is no single, simple, natural rate.

In a recent study (Jaynes & Cooke, 1987) of contemporary weathering rates in SE England (Fig. 4.14) 26 monitoring sites were established in a representative range of city, suburban, coastal and rural environments. At each site there was an SO$_2$ monitor (and occasionally other monitoring equipment, usually run by WSL or local authorities) and stone carousels. Each freely rotating carousel contained 12 stone tablets (50 mm $\times$ 50 mm $\times$ 10 mm), of which half were Portland stone and half were of Monk's Park stone—two common limestones with

very different physical properties. At each site, one carousel was fully exposed (and thus recorded the effects of both solution and sulphation), and another was sheltered beneath a cover (and thus eliminated the effect of run-off solution).

Three types of changes were recorded for these tablets at intervals over three years: weight loss,

Table 4.2 Some 'natural' rates of carboniferous limestone weathering in the UK

Area	Rate (μm year^{-1})	Method[a]
Bare limestone outcrop analysis		
NW Yorks	40	1, 2
C. Clare, Eire	9	1, 2
C. Clare, Eire	3–4	1, 2
S Wales	25–75	1, 2
Limestone catchment runoff analysis		
Peak District	75–83	
Mendips	40	
Mendips	23–88	
Mendips	81	

[a] Key to methods: 1—extrapolation from assumed glacial surface; 2—water sampling.

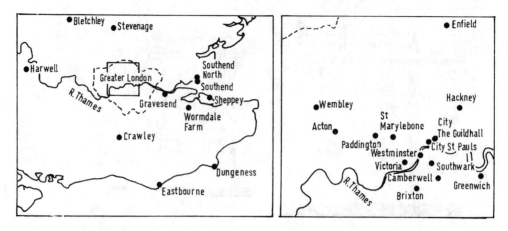

Fig. 4.14. Stone weathering monitor sites in south-east England (S. Jaynes, pers. comm.).

surface micromorphology and stone surface chemistry. The monitoring of surface roughness consistently revealed an increase of surface roughness over time, and a greater roughness increase on exposed than on sheltered samples.

Preliminary interpretation of results from this study reveals statistically significant differences in rates of weathering between urban and rural areas, but there are not simple gradients from areas of high to low pollution because the trends are locally interrupted by, for example, highs associated with coastal areas (where salt weathering is important) and unusually high local concentrations of SO_2. Within this broad conclusion, there appear to be several potentially important specific results. First, weight loss of exposed samples in Central London is only 25% greater than in rural (but also polluted) areas, although the SO_2 concentrations in Central London were three to four times greater than in the rural areas. This conclusion extends that from an earlier Building Research Establishment (BRE) study (Fig. 4.15), in which Portland stone examples exposed at Whitehall lost weight at about twice the rate of samples at Garston (near Watford), although SO_2 levels were five times higher in Whitehall in winter and over seven times higher in summer. In short, contemporary decay is less than directly proportional to ambient pollution levels.

Secondly, the weight loss of protected tablets was approximately 40% of that from exposed samples. Clearly the protected samples are being weathered, and the presence of sulphate suggests that sulphation is important. (Equally important, perhaps, is that the proportion of total weathering attributable to 'lower-carousel' processes is significantly lower in commercial and industrial centres than in out of town sites.)

4.5.4.2 Temporal variability

Evidence for historical rates of stone weathering can be obtained by various methods directly from buildings.

Two studies of St Paul's Cathedral illustrate the potential value of direct measurements of surface lowering. A lead plug index was used by Sharp *et al.* (1982). The coping stones on the balustrade at St Paul's have holes in them that were filled in *c.* 1717

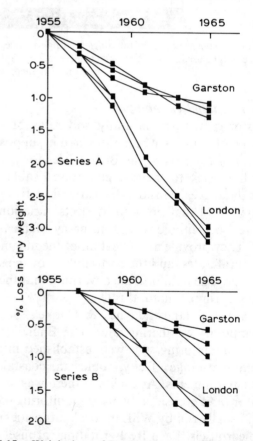

Fig. 4.15. Weight loss of Portland stone samples exposed at Garston (Herts) and Whitehall (Central London)(unpublished data, BRE).

with molten lead flush to the original stone surface. As a result of solution weathering in this most aggressive flow zone, after 262 years the 233 plugs stood proud of the stone surface by an average of 20·38 mm. The data reveal considerable variation around the building, with the maximum mean annual value of 83 μm year^{-1} in the south-west.

A second method is the use of a micro-erosion meter, which is placed on studs permanently fixed in the stone and records the elevation of the stone surface. Five years of measurements at the sites on the base of the balustrade (a location that is probably slightly less aggressive than the coping stone surface) gives a maximum annual rate, again in the south-west, of 490 μm year^{-1}.

A third approach involves the analysis of dated stone samples. At Wells Cathedral a unique collection of samples from the west front was obtained for each of several rock types (e.g. Doulting stone), that were emplaced at known, different times (e.g. fourteenth and nineteenth century) in identical micro-weathering environments. Petrological thin sections from such samples can reveal much about weathering history. For example, a Victorian sample of Doulting stone shows a stratigraphy in which a gypsum layer, heavily polluted near its surface with carbon, is overlain by a distinct deposit of calcium carbonate.

4.5.4.3 Damage on rain-sheltered and rain-washed surfaces

Much of the damage observable on historic buildings in the UK and elsewhere is the formation of friable crusts on relatively rain-sheltered surfaces of limestone and calcareous sandstone. Often these are blackened by incorporation of soot. When they spall off, this leads to dramatic loss of the original surface features.

It was established decades ago that the disfiguring crusts are rich in hydrated calcium sulphate, gypsum, formed as a result of dry deposition of sulphur dioxide into the pores of moist stone (Schaffer, 1932). Thicknesses of several millimetres have been produced in 50 years in atmospheres containing $\sim 300 \mu\text{g m}^{-3}$ SO$_2$. However, there is no reliable information on crust thickness as a function of ambient SO$_2$ concentration and time of wetness.

Luckat (1981) measured the weight gain of sandstone and limestone samples exposed in West Germany, in an attempt to relate damage to environmental variables in short-term experiments. Butlin *et al.* (1985) have looked at the weight loss of

similar limestone samples exposed in south-east England, after washing them to dissolve out reaction products. Both studies confirm the role of sulphur dioxide, though it is necessary also to invoke time of wetness (or some other variable) to explain UK data.

Although there has been no direct demonstration that the above weight loss or weight gain relates directly to eventual crust formation, this is not an unreasonable assumption. Luckat's (1981) data may therefore be used to indicate the relative effects of different SO$_2$ concentrations.

Luckat (1981) reported that in West German studies stone weight loss was a linear function of dry deposition of SO$_2$ measured by an 'immission rate monitoring apparatus' (IRMA). A wider study relating stone decay to SO$_2$ levels by the same passive method (IRMA) was covered at sites in Europe and America (NATO/CCMS study coordinated by NILU, Norway) and reached similar conclusions.

The correlation between immission rates and pollutant contents were investigated for two stone types (Baumberg and Muschelkalk). Limited success was achieved with SO$_2$ (correlation coefficient for linear regression around 0·7 for Baumberg, 0·5 for Muschelkalk). Better results were obtained with power functions (0·8, 0·6).

It was concluded that other parameters would have to be taken into account (e.g. meteorological measurements) before a more complete cause-effect analysis was possible. Measurements obtained with the IRMA device are dependent on approach velocity and to some extent on altitude (probably reflecting the aerodynamics on a given building). There were no extensive data available in these experiments to calibrate the device against standard methods of measurement of pollutant concentrations.

Extra meteorological data were provided by some of the participating countries for the years 1980–82, so that a more complete analysis of data from the 27 sites could be attempted. Correlations have been found between weight loss, SO$_2$ and the number of rain days.

The gypsum formation reaction

$$CaCO_3 + H_2SO_4 + H_2O \rightarrow$$
$$CaSO_4 \cdot 2H_2O + CO_2$$

is most likely with the sulphuric acid produced by oxidation of SO$_2$ and with moisture present in pores for relative humidities greater than 80%. Catalysts for this latter reaction can be found in soot and dust and it has been suggested that soot from oil is a better catalyst than soot from coal burning (Del Monte *et al.*, 1981).

Most investigations have dismissed the role of nitrogen oxides in stone decay because little nitrate is found in stone surfaces or in run-off water (Livingston, 1985). However, laboratory studies (Rosenberg & Grotta, 1980) suggest that NO_2 can enhance SO_2 pick-up without producing nitrates, and this may now be important in city centres. This requires further investigation in the field and in environmental chambers.

The rain-washed areas of limestone and sandstone buildings are often in better condition than the rain-sheltered areas (Schaffer, 1932; Blaeuer, 1985). However, some rain-washed areas of buildings are particularly susceptible to dissolution because of the microclimate and it is then important to know how dissolution is affected by acid pollutants in the air and in the rain.

A study of the weathering system at St Paul's Cathedral involved the recording of atmospheric variables in an automatic weather station, separate monitoring of pollutants (SO_2, NO_x, CO_2 and particulates at three levels), and run-off. Measurements were recorded on five alternate months in one year.

Water inputs to a defined catchment area were measured and run-off was collected and analysed as it passed across the building. A comparison of total hardness (Mg + Ca content) of rainfall for 24 rainfall events with the total hardness of run-off

from the same events led to a calculation based on a formula by Corbel (1959) of a solution rate of $220 \, \mu m \, year^{-1}$. This figure should be considered very approximate, however, because the method involves several assumptions. The run-off samples also included considerable quantities of sediment, approximately equivalent to $9.5 \, \mu m \, year^{-1}$, and it is clear that SO_4 and NO_3^- ions and the pH of run-off also increased, like total hardness, as it flowed across the surface (Table 4.3). Within the general statistics there are some interesting specific features. For example, on one thunderstorm day, pH was an exceptionally low 3·8, and this corresponded with the highest recorded total hardness (610 mg litre^{-1}) and the amount of particulates collected was also very high. This storm was also characterised by high concentrations of nitrate (18·3 mg litre^{-1}) about four times greater than average nitrate concentrations, while sulphate concentrations were lower. It would also have had the greatest rainfall intensity and probably the greatest wash-off of sediment (detached stone grains not converted to corrosion products).

Reddy *et al.* (1985) suggested that marble dissolution is proportional to hydrogen ion loading from the rain, but re-examination of their data shows that observed loss per millimetre of rain was independent of pH in their range (3·8 to 5). This would agree with the laboratory studies of Guidobaldi *et al.*, which show that calcium carbonate solution is insensitive to pH above pH 4 although it depends on surface roughness and rain intensity

Table 4.3 Data for five monitoring periods, St Paul's Cathedral, London

Period	pH (Mean) (a) Rainwater (b) Flowing water	SO_4 (mg litre^{-1})	NO_3 (mg litre^{-1})	Total hardness (mg litre^{-1})
1 (16 October–12 November 1980)	(a) 4·79 (b) 6·59	21·8 23·6	3·95 4·80	4·83 154
2 (12 January–8 February 1981)	(a) 4·56 (b) 5·95	15·8 17·2	4·46 5·06	7·0 211
3 (9 March–5 April 1981)	(a) 4·88 (b) 6·41	17·2 19·9	4·6 5·4	15·2 96·8
4 (4–31 May 1981)	(a) 4·61 (b) 6·31	14·3 18·1	4·6 7·2	13·5 119·5
5 (27 June–24 July 1981) Day 1	(a) 3·8 (b) 6·0	3·8 4·4	18·3 19·5	24 610
Day 2	(a) 4·3 (b) 6·9	13·1 14·6	4·4 4·8	0 171

Source: Butlin *et al.* (1985).

(Guidobaldi, 1981; Guidobaldi & Mecchi, 1985). More research is required in order to establish the significance of rain episodes in which the pH falls below 4, and also on the relative effects of dry and wet deposition on stone decay.

4.5.4.4 Lichens, air pollution and weathering
The role of lichens, algae and bacteria in stone decay, and methods of destruction/prevention of organic growths, have been under study for some time at BRE. Work has recently been started on the role that a surface coverage of organisms, and especially lichens, plays in stone weathering and how this role is influenced by air pollution. Lichens are known to influence stone surface weathering (in some situations encouraging weathering, in others protecting the surface from rainfall effects) and are also sensitive indicators of atmospheric pollution. Lichens are long-lived organisms which can be dated relatively easily and the study of lichen community structure and change over time may provide evidence on changes in pollution level. Detailed studies of lichen communities at Wells Cathedral and other sites are currently being carried out in order to assess the role played by lichens in processes of decay at the present time and their use as indicators of building stone decay and air pollution histories.

A German study on Cologne Cathedral has concluded that bacteria in the stone convert pollutants into nitric acid which causes the decay (Anon., 1986).

4.5.4.5 Effect of particulates
Some recent studies on the effect of particulates on stone by Camuffo *et al.* (1982) and Del Monte *et al.* (1981) have examined the thesis that only sheltered stone is affected by sulphur compounds due to generation of low pH via condensation on black particulate residues of oil and coal combustion. Camuffo *et al.* believe that there is no direct effect of wet acid precipitation on stone nor of recrystallisation processes within the stone as part of the decay processes. These views are somewhat controversial and can only be tested by more experimentation.

4.5.4.6 Other materials
There is a possibility that glass, organic materials (some of which are 'filled' with inorganic oxides), brickwork, cement and mortars may be affected by pollutants. An important feature to be considered

in each case is whether the natural weathering characteristics are affected by pollutant species solely or in combination with other meteorological variables.

(a) *Brickwork.* Bricks vary in durability depending upon the pore-size distribution, chemical and mechanical properties. Brickwork can be affected indirectly by salts which move from the brick into the mortar joints causing expansive reactions. Salts in the bricks themselves cause the self-evident efflorescence. Little is known, however, about the likely effect of acid deposition on bricks or brickwork and the extent to which weathering characteristics (e.g. frost susceptibility) would be affected.

(b) *Concrete.* The effects of pollutants on bulk concrete are minor compared to the effects of reinforcement corrosion considered in section 4.4.3.7.

It is unlikely that SO_2 has a major role in reducing the surface alkalinity of concrete compared to the well-known effects from CO_2, which is present in the atmosphere in much higher concentrations. Concrete itself can react with SO_2 to form ettringite in an expansive reaction.

Further studies would be necessary to investigate pollutant effect on concrete surfaces, but a degree of surface retreat may be tolerated with loss of serviceability only occurring if the aggregate particles became detached.

(c) *Glass.* Although modern glazing is highly durable, there is understandably much concern over the possible attack by pollutants, in particular on mediaeval painted glass. Such glasses used different raw materials, which contained higher proportions of potash and lime than modern soda-lime glasses. Corrosion phenomena on painted glass are extremely complex and systematic relationships are difficult to recognise. It is believed, however, that pollution in recent years has accelerated the corrosion of susceptible glasses. Leached alkali products from the glass itself can produce an alkaline solution on the surface of the glass which, if of pH 9, can lead to rapid dissolution of the silica network in the glass and may react with acid gases to form porous salt crusts. There is research work currently underway in which the corrosion mechanisms are being further examined and preventative/restorative methods under development. It should be said, however, that current evidence points to moisture being important and to only a secondary role of SO_2, in that it may produce hygroscopic reaction products which act as a wet poultice on the surface of the glass—thus encouraging further breakdown of the silica network. There is no evidence that SO_2 initiates the breakdown process, although it accelerates the overall reaction.

4.6 TAKING ART INDOORS

Buildings and monuments are more robust than people. Where measures have been taken to control atmospheric pollution, the motive has almost invariably been to protect the comfort and health of *people*. Of course, as a bonus, buildings have also benefited but, due to the 'memory' effect referred to earlier, the reduction in rate of corrosion has not been as large as might have been hoped.

Structures have thus, for most of history, been left to take care of themselves, and this is likely to remain the case whatever the outcome of the present 'Acid Rain' debate, of which this report is a part. In fact, with a few notable exceptions, the main fabric of buildings has not been the main source of worry. It is their decoration—the carvings and free-standing sculpture—which has given rise to most cause for concern.

The much-criticised practice of the Imperial Powers of 'plundering' the monuments of the conquered, and carrying off works of art to swell their national collections, has in fact done much to preserve what might otherwise have been lost due to the ravages of corrosion. The Parthenon Frieze, which forms the main part of the Elgin 'Marbles' at the British Museum, is without question in much better shape today than had it been left in place, and exposed to the attention of even less scrupulous 'collectors' or 'restorers' or suffered the corrosive attack of the polluted Athenian atmosphere.

At a somewhat later stage in history, the practice began of substituting copies of the works of art in the original locations. This can be said to have started in 1871 when the British Museum Authorities, perhaps feeling pangs of conscience, installed on the Porch of the Erechtheum in Athens a concrete copy of the Caryatid which Lord Elgin had removed over half a century earlier, and which formed part of the Museum's permanent collection. Quite recently the Greek Conservators themselves have emulated Elgin and the British Museum's action and brought indoors the remaining Caryatids, and substituted copies on the Temple itself.

There are many other examples of 'substitution': in 1873 Michelangelo's 'David' was removed to a museum from its original open-air position in the Palazzo Vecchio in Florence, and was replaced by a rather poor copy. In the same city, Donatello's statue of St George has been removed from its niche on the outside of the Orsanmichele and is now in the Borgello Museum; again a copy has been placed in the original position. No doubt the Florentines will soon be forced to take similar action with Ghiberti's bronze doors of the Baptistry, and Cellini's bronze sculpture 'Perseus'. In Venice the four bronze horses of St Mark's have been moved under cover and copies substituted, and the bronze equestrian statue of Marcus Aurelius in Rome has also been removed to the protection of a museum. In Paris, Carpeaux's masterpiece, the stone group 'The Dance' has been removed from the facade of L'Opéra to the Louvre.

Even when sculptures, and other artefacts, are placed inside museums, their troubles may not be over—the museum atmosphere itself can be very damaging. A sixteenth century limestone effigy of Christ recovered in good condition from damp earth beneath the ruins of Mercer Hall in London, exhibited spalling of its surface layers after only 20 years in an air-conditioned room in the Guildhall Museum. There is also the famous case of the iron cannon salvaged in 1953 from the wreck of the *Riksapplet*, which foundered in 1676. In spite of spending nearly 300 years on the sea bed, the decoration on the cannon was still in excellent condition when it was first brought to the surface. However, though the cannon was given an anti-rust treatment, corrosion occurred in the museum atmosphere and after 18 years of this attack the decoration had almost completely disappeared. New concrete museums may present a particular hazard as the concrete generates a fine alkaline dust which can pass through air filters and be very damaging. Neither are the museums themselves safe from 'natural' disasters—the devastation caused by the floods in Florence in 1966 is perhaps the worst example of recent times. Within the past year our own Victoria and Albert Museum has had works of art damaged by the flooding of a basement.

Of course buildings and structures are often too large to be brought indoors, and other methods of protection may have to be employed—there has been a proposal, for example, to protect the Albert Memorial by constructing around it a giant pyramid of glass.

In some cases the work of art has had to be removed from sight of the public altogether. The caves at Lascaux, for example, with their wonderful palaeolithic paintings, have been closed since 1963, and it has recently been announced that the public will be excluded from viewing Leonardo da Vinci's 'Last Supper' for a period of at least five years.

In other cases where it was not practical to remove the work of art—say some Gothic carving on a cathedral—then plaster casts have been taken, and exhibited in a museum. We are particularly fortunate in this country in having, in the Cast Courts at the V and A, one of the world's finest collections of casts.

The principal motive in the past for producing casts has been for the education and enjoyment of a public who might not be able to travel to see the originals. It is not necessary, for example, to travel to Italy to view a number of Michelangelo's and Donatello's finest sculptures—they are reproduced as plaster casts at the V and A. However, taking casts can play a most valuable additional role; that is, in accurately recording for posterity the image of a work of art as it existed at some point in time before the original had been damaged or destroyed. Perhaps the best illustration of this is the fact that the famous fifteenth century Lübeck relief of Christ washing the feet of the Apostles has been destroyed but, by great good fortune, a cast of it is in the V and A collection.

This is a unique case; more typical is the cast which shows details no longer to be seen on the original due, for example, to incompetent attempts at restoration—a case in point is the reliefs from the S. Maria dei Miracoli in Brescia. Other original pieces suffer the ravages of corrosive attack from polluted atmospheres. In the brochure on the V and A collection photographs are reproduced showing the fine detail on a hundred-year-old cast of a tympanum which has by now been badly eroded on the original. Also the detail on the huge cast of Trajan's Column, which dominates the collection, is already superior to that on the original in Rome. This litany is, however, brought to conclusion with an account of a failure. A cast was taken in the middle of the last century of a fine twelfth-century carved doorway from a church at Shobden in Herefordshire; subsequently the original was destroyed by weathering. Tragically the cast too was lost in the 1930s when it was broken up simply to make more space available in the museum. This is a mistake which must not be repeated.

Taking plaster casts is not the only method of reproduction: the V and A has also an unrivalled collection of electrotype reproductions of many of the world's greatest examples of the goldsmith's art.

A beautiful copy of a pair of Ghiberti's bronze doors for the Baptistry of Florence provides a good insurance against any possible damage to the original, which remains exposed to the elements.

The practice of taking casts reached its height during the Victorian era, and has been out of fashion for most of this century. This is to be regretted. Although it is true to say that few of the greatest works of art are still exposed to the atmosphere, many irreplaceable carvings in stone still are, and the taking of casts, if revived, would provide a record of much that is of value before weathering takes its final toll.

4.7 CONCLUSIONS

The available evidence shows clearly that the rates of corrosion of the metals commonly used in construction increase markedly with the concentration of sulphur dioxide and with time of wetness. For other materials, effects of atmospheric pollution seem likely to be less serious. The effects of pollutants other than sulphur dioxide are less well understood. SO_2 levels have certainly fallen considerably in the last 30 years in the urban areas where most building materials are concentrated, and there is some evidence that rates of corrosion have fallen in line with this reduction. It is possible that NO_x is deleterious mainly in combination with SO_2, but more information is needed on this point.

Surveys suggest that repainting cycles are shorter in more polluted atmospheres, but it is not clear that this is related to the economic life of the paint rather than to aesthetic considerations or prestige: limited evidence suggests that the life of paint films is not greatly affected by present pollutant levels.

Calculations of the cost of corrosion are difficult, since the cost includes extra expenditure on protective measures, maintenance and non-availability, as well as replacement. Many associated costs are difficult to quantify, since, for example, the failure of a fastener may entail the dismantling, if not the total replacement, of major building components at a cost far exceeding that of the failed item. Many such failures are due to errors in design, or in building practice, or selection of materials, but their consequences become far more serious as a result of corrosion accelerated by pollutants.

There is no doubt that acid deposition can affect building materials, especially those containing calcium carbonate. The existence of gypsum on and in the surfaces of stone has been known for a long time. The current difficulties in determining the causes of the decay include:

— the dearth of measurements on actual rates of decay of building surfaces;
— disaggregating the natural rates of weathering and decay process from those associated with polluted atmospheres;
— differentiating between the effects of dry and wet deposition;
— determining whether current levels of decay are different from those when pollution levels were higher;
— disassociating the longer term pollution effects from the shorter term.

The indications so far are that there is no evidence that in-situ decay rates (for stone) are decreasing, that effects are highly directional and are associated with meteorological parameters and that there is no simple correlation between SO_2 concentrations and rates of decay and no direct effect of NO_x (although there may be an indirect effect). There are discernible differences between the effects of SO_2 in urban and industrial areas compared with rural areas, and there can be high levels of dissolution of, for example, Ca and Mg in single rainfall episodes. Current research is pursuing many of the as-yet unsolved aspects of pollution effects.

As far as metallic materials are concerned, it is probably true to say that reasonably practical and economical means are available for protecting those used in the construction industry against present-day levels of atmospheric pollution, and that the main effect of pollutants may be to exacerbate problems caused by unsatisfactory design or building practice.

ACKNOWLEDGEMENT

Some of the experts who compiled this section also contributed to the Report of the Building Effects Review Group being prepared concurrently for the DoE and this has led to some textural overlap between this section and parts of the BERG report.

REFERENCES

Anon. (1986). *Chemistry in Britain* (Dec. 1986), 1072.

BALL, D. J. & HUME, R. (1983). The relative importance of vehicular and domestic emissions of dark smoke in Greater London in the mid 1970s, the significance of smoke shade measurements and an explanation of the relationship of smoke shade to gravimetric measurements of particulate. *Atmos. Environ.*, **17**, 169–81.

BARRETT, C. F. *et al.* (1983). *Acid deposition in the United Kingdom, First report of the United Kingdom Review Group on Acid Rain (RGAR)*. Warren Spring Laboratory, Stevenage.

BARRETT, C. F. *et al.* (1987). *Acid deposition in the United Kingdom 1981–85, Second report of the United Kingdom Review Group on Acid Rain (RGAR)*. Warren Spring Laboratory, Stevenage.

BARTON, K. & BARTONOVA, Z. (1969). Über den beschleunigenden Einfluss von Schwefeldioxid und Wasser auf die atmosphärische Korrosion von verrosteten Eisen. *Werkstoffe u. Korros.*, **20**, 216.

BAWDEN, R. J. & FERGUSON, J. M. (1987). Trends in materials degradation rates in the UK. In: *Proceedings of UK Corrosion '87, Brighton, 26–28 October*, pp. 383–402.

BETTELHEIM, J. & LITTLER, A. (1979). Historical trends of sulphur oxide emissions in Europe since 1865. CEGB Report PL-GS/E/1/79.

BISRA (1938). Fifth report of the corrosion committee. Iron and Steel Institute, Special Report No. 21, pp. 63–5.

BLAEUER, C. (1985). Weathering of Bernese sandstones. In: *Proc. 5th Int. Congr. Deterioration and Conservation of Stone, Lausanne, 25–27 Sept. 1985*, ed. G. Félix, Presses Polytechniques Romandes, p. 381.

BRIMBLECOMBE, P. (1977). London air pollution 1500–1900. *Atmos. Environ.*, **11**, 1157–62.

British Standards Institution (1977). *Code of Practice for Protective Coating of Iron and Steel Structures*, BS5493.

British Standards Institution (1982). *Code of Practice for Painting of Buildings*, BS6150.

BUTLIN, R. A., COOKE, R. U., JAYNES, S. & SHARP, A. (1985). *5th Int. Congr. Deterioration and Conservation of Stone, Lausanne, 25–27 Sept. 1985*, ed. G. Félix, Presses Polytechniques Romandes.

CAMUFFO, D., DEL MONTE, M., SABBIONI, C. & VITTORI, O. (1982). *Atmos. Environ.*, **16**, 2253.

CHAMBERLAIN, A. C., GARLAND, J. A. & WELLS, A. C. (1984). Transport of gases and particles to surfaces with widely spaced roughness elements. *Boundary Layer Meteorol.*, **29**, 343–60.

CORBEL, J. (1959). *Ann. Geogr.*, **68**, 97.

CORDNER, R. J. *et al.* (1984). In: *Proc. Int. Congr. Metal Corrosion, Toronto*, 206.

CORVO-PEREZ, F. (1984). In: *Proc. Corrosion (Houston)*, **40**, 170.

DEL MONTE, M., SABBIONI, C. & VITTORI, O. (1981). Airborne carbon particles and marble deterioration. *Atmosph. Environ.*, **15**, 645.

DERWENT, R. G. (1986a). The nitrogen budget for the UK and NW Europe. Energy Technology Support Unit Report R-37, ETSU, AERE, Harwell, UK.

DERWENT, R. G. (1986b). Atmospheric ozone and its precursors. Energy Technology Support Unit Report R-38, ETSU, AERE, Harwell, UK.

EVANS, U. R. (1960). *Corrosion and Oxidation of Metals*. Arnold, London.

FELIU, S. *et al.* (1984). *Rev. Iberoam. Corros. Prot.*, **15**, 11.

FINK, W. *et al.* (1971). *Technical-Economic Evaluation of Air Pollution Corrosion Costs on Metals in the USA*. Battelle, Columbus, PB198453.

GARLAND, J. A. (1977). The dry deposition of sulphur dioxide to land and water surfaces. *Proc. Roy. Soc. Lond.*, **A354**, 245–68.

GONZALEZ, J. A. & BUSTIDAS, J. M. (1982). *Rev. Iberoam. Corros. Prot.*, **13**, 19.

GRAEDEL, T. E. (1986). Corrosion-related aspects of the chemistry and frequency of occurrence of precipitation. *J. Elec. Soc.*, **133**, 2476.

GUIDOBALDI, F. (1981). Acid rain and corrosion of marble:

an experimental approach. In: *Proc. 2nd Int. Symp. on Conservation of Stone*, Bologna, p. 483.

GUIDOBALDI, F. & MECCHI, A. M. (1985). Corrosion of marble by rain: The influence of surface roughness, rain intensity and additional washing. In: *Proc. 5th Int. Congr. Deterioration and Conservation of Stone, Lausanne, 25–27 Sept. 1985*, ed. G. Félix, Presses Polytechniques Romandes, p. 467.

GULLMAN, J. & SWARTLING, P. (1983). *Korroze. Ochr. Mater.*, **27**, 52.

GUTTMAN, H. (1968). Effects of atmospheric factors on the corrosion of rolled zinc. ASTM STP 435, American Society for Testing and Materials, Philadelphia, p. 223.

GUTTMAN, H. & SEREDA, P. J. (1968). ASTM STP 435, American Society for Testing and Materials, Philadelphia, pp. 326–59.

HAAGENRUD, S., HENRIKSEN, J. F. & GRAM, F. (1985). Paper No. 138 presented at Electrochemical Society Fall Meeting. Corrosion Effects of Acid Deposition, 13–18 October, Las Vegas.

HARRIS, J. E. (1984). Oxidation induced deformation and fracture. In: *Proc. 6th Int. Conf. Fracture, New Delhi, India, 4–10 December 1984*.

HAYNIE, F. H. & UPHAM, J. B. (1970). Effects of atmospheric sulphur dioxide on the corrosion of zinc. *Materials Performance*, **9**, 35.

HIMI, Y. (1981). *Kanagawa-Ken Taiki Osen Chosa Kenkyu Hokoku*, **23**, 167.

HUDSON, R. M. (1986). The effect of environmental conditions on the performance of steel in buildings. In: *Proc. Conf. 'Atmospheric Corrosion and Weathering', 3 December, London, MAFF*.

HUDSON, J. C. & STANNERS, J. F. (1953). The effect of climate and atmospheric pollution on corrosion. *J. Appl. Chem.*, **3**, 86.

JACKSON, R. L., GIBBON, R. R., FERGUSON, J. M. & LEWIS, K. G. (1987). The condition of the supergrid and a strategy for refurbishment. In: *Proc. IEE Conf. 'Revitalising Transmission and Distribution Systems', London, 25–27 February 1987*.

JAYNES, S. & COOKE, R. U. (1987). Stone weathering in South East England, *Atmos. Environ.*, **21**, 1601–22.

JOHANSSON, L-G. (1985). A laboratory study of the influence of NO_2 and SO_2 on the atmospheric corrosion of steel, copper, zinc and aluminium. Paper No. 142 presented at Electrochemical Society Fall Meeting, Corrosion Effects of Acid Deposition, 13–18 October, Las Vegas.

JOHNSON, K. E. (1982). Airborne contaminants and the pitting of stainless steels in the atmosphere, *Corr. Sci.*, **22**, 175.

JORG, F., SCHMITT, D. & ZIEGAHN, K.-F. (1985). Materials damage due to air pollution, Parts I and II. Fraunhofer Institute Report No. UBA-FB 84-106 08 010.

KAMAYA, T. *et al.* (1981). *Nagasacki-Ken Eisci Rogai Renyuscho Ho*, **23**, 25.

KEDDIE, A. W. C. *et al.* (1977) The high pollution episode in London, December 1975. Report LR 263(AP), Warren Spring Laboratory, Stevenage.

KNOTKOVA-CERMAKOVA, D. & MAREK, V. (1976). *Freiberg Forsch.*, **B189**, 59.

KNOTKOVA-CERMAKOVA, D., BOSEK, B. & VLCHOVA, J. (1974). Corrosion aggressivity of model regions of Czechoslovakia. ASTM STP 558, American Society for Testing and Materials, Philadelphia, p. 52.

LEARY, E. (1983). *The Building Limestones of the British Isles*. HMSO, London.

LIPFERT, F. W., BENARIE, M. & AAUM, M. L. (1985). Paper No. 137 presented at Electrochemical Society Fall Meeting, Corrosion Effects of Acid Deposition, 13–18 October, Las Vegas.

LISS, P. S. (1971). Exchange of SO_2 between the atmosphere and natural waters, *Nature, Lond.*, **233**, 327–9.

LIVINGSTON, R. A. (1985). The role of nitrogen oxides in the deterioration of carbonate stone, *Proc. 5th Int. Congr. Deterioration and Conservation of Stone, Lausanne, 25–27 Sept. 1985*, ed. G. Félix, Presses Polytechniques Romandes, p. 509.

LUCKAT, S. (1981). *Staub-Reinhalt der Luft*, **41**, 128.

MAFF (1982). Atmospheric corrosion of zinc (map). HMSO, London.

MAFF (1986). *UK Atmospheric Corrosivity Values*. MAFF, London.

NRIAGU J. O. (1978). *Sulfur in the Environment, Pt II*. J. Wiley, New York.

PAYRISSAT, M. & BEILKE, S. (1975). Laboratory measurements of the uptake of sulphur dioxide by different European sites. *Atmos. Environ.*, **9**, 211–17.

POURBAIX, M. (1982). *Atmospheric Corrosion*, ed. W. H. Ailor. John Wiley, New York, p. 167.

REDDY, M. M., SHERWOOD, S. & DOE, B. (1985). Limestone and marble dissolution by acid rain. *Proc. 5th Int. Congr. Deterioration and Conservation of Stone, Lausanne, 25–27 Sept. 1985*, ed. G. Félix, Presses Polytechniques Romandes.

ROGERS, F. S. M. (1984). A revised calculation of gaseous emissions from UK motor vehicles. Report LR 508(AP), Warren Spring Laboratory, Stevenage.

ROSENBERG, H. S. & GROTTA, H. M. (1980). *Environ. Sci. Technol.*, **14**, 470–2.

ROZENFELD, I. L. (1972). *Atmospheric Corrosion* (English translation). NACE, Houston.

SAUNDERS, K. G. (1982). Paper presented at 49th Annu. Conf. Nat. Soc. For Clean Air.

SCHAFFER, R. J. (1932). The weathering of natural building stones. *BRE Special Report*, No. 18, HMSO, London.

SCHICKORR, G. (1963). Über den Mechanisms des atmosphärische Rostens des Eisens. *Werkstoffe u. Korros.*, **14**, 69–80.

SHARP, A. D. *et al.* (1982). Weathering of the balustrade on St Paul's Cathedral, London. *Earth Surface Processes and Landforms*, **7**, 387.

SHAW, T. R. (1978). ASTM STP 646, American Society for Testing and Materials, Philadelphia, pp. 204–15.

SKOULIKIDIS, T. N. (1982). In: *Atmospheric Corrosion*, ed. W. H. Ailor. Wiley Interscience, New York, pp. 809–10.

SPEDDING, D. J. (1969). Sulphur dioxide uptake by limestone. *Atmos. Environ.*, **3**, 683–5.

SPENCE, J. W., EDNEY, E. O., STILES, D. C. & HAYNIE, F. H. (1985). Paper No. 133 presented at Electrochemical Society Fall Meeting, Corrosion Effects of Acid Deposition, Las Vegas, 13–18 October.

TULKA, J. & SCHATTAUEROVA, A. (1982). *Korroze. Ochr. Mater.*, **26**, 68.

Umweltbundesamt (1980). *Luftverschmutzung durch Schwefeldioxid*. Berlin.

UNECE (1984). Air-borne sulphur pollution—effects and control. ENV/IEB R16, February 1984.

VERNON, W. H. J. (1935). A laboratory study of the atmospheric corrosion of metals. Part III: The secondary product or rust (the influence of sulphur dioxide, carbon dioxide and suspended particles on the rusting of iron). *Trans. Farad. Soc.*, **31**, 1678.

WALTON, J. R., JOHNSON, J. B. & WOOD, G. C. (1982a). Atmospheric corrosion initiation by sulphur dioxide and particulate matter, I, *Brit. Corr. J.*, **17**, 59.

WALTON, J. R., JOHNSON, J. B. & WOOD, G. C. (1982b). Atmospheric corrosion initiation by sulphur dioxide and particulate matter, II. *Brit. Corr. J.*, **17**, 65.

WEATHERLEY, M-L. P. M., GOORIAH, B. D. & CHARNOCK, J. (1975). Fuel consumption, smoke and sulphur dioxide emissions and concentrations, and grit and dust deposition in the UK, up to 1973/4. Report LR 214(AP), Warren Spring Laboratory, Stevenage.

WSL (1967). *The Investigation of Atmospheric Pollution 1958–1966*, 32nd Report. HMSO, London.

WSL (1972). *National Survey of Air Pollution 1961–71*, Vol. I. HMSO, London.

YASUKAWA, S. *et al.* (1980). *Boshoku Gijutsu*, **29**, 609.

Section 5

Control and Remedial Strategies

Cyril Davies

Operational Research Executive, British Coal, Harrow, Middlesex

This paper presents the work of a sub-group of the
Watt Committee working group on Air Pollution,
Acid Rain and the Environment.

Membership of Sub-group

C. J. Davies (Chairman)

W. V. C. Batstone
A. J. Clarke
Dr M. J. Cooke
Dr D. Cope
P. Dacey
M. J. Flux
Dr G. D. Howells
P. Jones
Prof K. Mellanby
B. Mould
Dr J. Skea
P. F. Weatherill
Dr J. H. Weaving
Dr M. Williams
M. Woodfield

5.1 SCOPE AND OBJECTIVES

An effective strategy for the control and abatement of the effects of fossil fuel use depends upon a good understanding of three sets of factors:

—scientific: how are pollutants transformed and transported, and what are the damage mechanisms?
—engineering/technological: how can pollutant emissions be reduced? and
—economic/legislative: what are the costs of emission reduction and what mechanisms are available to achieve reductions most effectively?

This section comments on the second and third of these factors. The objective is to present the technological options for control and abatement with their costs and to indicate, in the light of likely developments in UK energy usage over the next decades, the scale of implementation of these technologies that would be necessary to reduce UK emissions to some target level. The material presented updates and expands our earlier Watt Committee report of August 1984. In particular, it reflects the growing consensus over the last few years that there are other emissions in addition to oxides of sulphur that play a major role in the complex phenomena of acid rain. We have put more emphasis, therefore, on emissions from mobile sources, and on technologies that reduce nitrogen, as well as sulphur oxide emissions from thermal plant.

The range of issues relevant to designing a policy on atmospheric pollution is considered from various viewpoints. In general the expertise and experience of members of the Working Group extended over no more than a few of these. The section is, therefore, a series of only loosely-linked papers, each of which has been drafted by one or two individuals. However, as a Group we have sought to ensure overall balance in the material presented and in any conclusions that can be drawn from available evidence.

The first paper looks at the international context in which any UK legislation or policy will inevitably be set. It discusses the two major current initiatives, from the European Commission and from the United Nations Economic Commission for Europe.

Following this is a paper indicating the ways emissions of sulphur dioxide, nitrogen oxides and hydrocarbons might evolve if there were no further legislation in the UK. This is based upon a sectorial analysis of emissions and takes account of the major

identifiable uncertainties. The paper considers the role conservation and non-fossil energy sources could play in reducing emissions and so allows the need for emission-reducing techniques to estimate to reach alternative emission levels. The analysis is based upon work of the Science Policy Research Unit.

Three papers follow which deal with the technical and engineering aspects of emissions reduction. The first reviews control from stationary sources. Technologies for sulphur and nitrogen oxides removal are described, covering plant sizes from central power generation to general industrial scale plant. The second of the three is concerned with the state of play for emissions from mobile sources. It explains the basic emission characteristics of internal combustion engines and goes on to describe the several approaches to controlling these emissions. Following this is a paper covering monitoring aspects — the requirements for monitoring under the European Community proposals, the available technologies and the likely costs.

The implications of environmental legislation on industry are considered next. In the UK most debate has focussed upon the power-generating sector for which the facts are relatively easily available. However, potential legislation could well affect plant of industrial scale. The consequences are much more difficult to assess and the paper looks at some of the key issues and uncertainties.

We move next to a paper describing experiences of liming lakes. This is the most important example of actions directly ameliorating the effects of acid rain. It is likely to have a part to play in enabling acidified lakes to re-establish self-supporting fish populations, whether levels of acid-producing emissions are reduced or not. Promising results need not be seen, therefore, as lessening the need for emission reduction; rather liming of lakes could be seen as complementary. There is some discussion of liming lakes in Section 3.2.7.2 of this report. A concern in the present context has been to present information which addresses prospective cost-effectiveness and practical factors.

The final paper of the section discusses some of the principles necessary to design a control and abatement policy which is efficient in economic terms. It is clear that, given the scientific uncertainties, any policy must involve political judgement as much as scientific calculation. However, approaches are being developed to provide support in exercising judgments, and in translating these into effective policies.

Where it seems safe to do so, conclusions are presented for the separate papers. We have not sought to provide conclusions for the section as a whole which go beyond these.

5.2 EUROPEAN VIEW AND LEGISLATIVE PROSPECTS

UK legislation on atmospheric emissions must be viewed against the backcloth of developments in Europe. This paper concentrates on the two principal current initiatives: first, the United Nations Economic Commission for Europe Convention on Long Range Transboundary Air Pollution and, secondly, the European Economic Community proposed Directive on Limiting Emissions from Large Combustion Plant. It relates to the situation at the end of 1985 when this paper was originally presented. However, despite much discussion and activity, the situation remains essentially unchanged 18 months later.

5.2.1 UNECE Convention and 30% Protocol

The UNECE Convention is an international treaty signed by 35 nations. It includes general undertakings on emission reduction and exchange of information. It has recently given rise to a second, separate agreement between some of its signatories known as the 30% Protocol or, more popularly, as the "30% Club". This includes those signatories who have committed themselves to reducing sulphur dioxide emissions (or, alternatively, transboundary fluxes of sulphur dioxide) by 30% in 1993, compared to their corresponding emissions or fluxes in 1980. It is important to note that the 30% Protocol stands on its own as an international agreement, separate from the UNECE Convention; signatories of the latter are under no formal obligation to sign the former. The UK, though a signatory of the UNECE Convention, has not joined the 30% Club.

Table 5.2.1 shows all the 35 signatory nations to the Convention in two columns: on the left, the 21 nations who have also signed the 30% Protocol, and on the right, the 14 signatories who have not made this commitment. The latter include the UK, but we are not alone. The uncommitted members, besides the UK, include Spain and Poland, both of whose sulphur dioxide emissions are broadly comparable to the current levels from the UK. (In passing, too much should not be made of the fine differences in the published estimates of international sulphur dioxide emissions. The accuracy of the figures is highly speculative and any figures in the range, of say, 2·5–3·5 million tonnes per annum can be considered comparable.)

Another outstanding non-member of the 30% Club is the USA, a fact that is frequently submerged by concentrating on the European context. The reasons that the USA have given for not signing the Protocol are a close echo of those put forward by the UK, that is, that not enough is known about the causes and effects of damage to lakes and forests to be sure that an arbitrary reduction of 30% will have any beneficial effect. Both countries call first for a greater effort to understand these problems, which is unsatisfactory to those countries and individuals who consider that there is sufficient evidence of environmental damage to required urgent action.

Among the members of the 30% Club, the Soviet Union, probably speaking on behalf of all the seven Eastern Bloc members at Geneva, made two things quite clear. First, that they view their commitment as reducing transboundary fluxes by 30% only between themselves and other signatories to the Convention, that is, only over their western borders and not to the World as a whole. Secondly, like France, they expect to achieve the 30% reduction by substantially increasing nuclear generation and by converting existing boilers to low-sulphur fuels, which will undoubtedly turn out to be natural gas. It seems clear that these countries are firmly convinced that these changes in basic energy pattern are

Table 5.2.1 UNECE Transboundary Convention

30% Protocol members	Uncommitted members
Austria	European Commission
Belgium	Greece
Bulgaria	Holy See
Byelorussian SSR	Iceland
Canada	Ireland
Czechoslovakia	Poland
Denmark	Portugal
Finland	Romania
France	San Marino
German Democratic Republic	Spain
Federal Republic of Germany	Turkey
Hungary	*United Kingdom*
Italy	United States of America
Liechtenstein	Yugoslavia
Luxembourg	
Netherlands	
Norway	
Sweden	
Switzerland	
Ukranian SSR	
USSR	

economically justifiable in their own right. Whilst not a member of the Protocol, the UK has announced a 'policy intention' to reduce both oxides of sulphur and of nitrogen by 30% of 1980 levels by the late 1990s.

5.2.2 EEC proposed directive on large combustion plant[†]

The proposed European Community Directive includes both a requirement for countries to reduce overall emissions by 1995 compared with 1980, and emission standards for new plant. In contrast to the UNECE Protocol, the percentage reductions required relate only to emissions from plant over 50 MW rather than to all plant. (It is expected that Community proposals for small plant will follow at a later stage.) The proposals relate to sulphur dioxide, nitrogen oxide and particulate emissions. Table 5.2.2 summarises the proposals.

There is some fluidity in the positions taken by individual member states. For illustration, Table 5.2.3 shows the position that the individual Member States are *believed* to have taken on the proposed

Directive, as it appeared at the end of 1985. The negotiations are not fully reported, so there is an element of hear-say in this matter. It is clear from pronouncements made outside the proceedings that four countries have entered a general reservation against the whole draft of the Directive. It is understood that some, if not all, of the remaining countries have reservations on important points of detail in the draft. Greece and Ireland, as small emitters, argue with reason that the Directive will be extremely harsh in its economic impact, without securing any material improvement to the European environment. Italy, on the other hand, is a major emitter with problems of implementation similar to those of the UK.

In January 1986, two other countries became full Member States in the Community: Spain and Portugal. Their formal position on the Directive has not, at the time of writing, been stated but it seems very unlikely that either will be prepared to accept the draft. Neither have joined the 30% Club; in the negotiations that led up to the Protocol, Spain tried hard to have wording adopted that would limit the signatories merely to preventing an increase in

Table 5.2.2 Proposed emission limit values in waste gases under standard conditions

Type of fuel	Plant size (MW(th))	Dust	Sulphur dioxide[a] Emission limit values for: (mg m^{-3})		Oxides of nitrogen[b]	
			From 1 Jan. 1985	After 31 Dec. 1995	From 1 Jan. 1985	After 31 Dec. 1995
Solid	>300 300–100 <100	<50	<400[c] <1200 <2000	<250[d] <1200[d] <2000[d]	<650[c] <800 / <800 as a rule but <1300 for pulverised hard coal firing with extration of fused ash	<200[d] <400 as a rule[d] but <800 for pulverised hard coal firing with extraction of fused ash
Liquid	>300 300–100 <100	<50	<400 <1700 <1700	<250 <1700 <1700	<450	<150
Gaseous	>300 300–100 <100	<5 as a rule but <10 for blast furnace gas and <50 for gases produced by the steel industry which can be used elsewhere	<35 as a rule but <100 for coke oven gas and <5 for liquified gas		<350	<100

[a] Account should be taken of the proportion of sulphur trioxide in the waste gas.
[b] Expressed in terms of nitrogen dioxide. Nitrogen dioxide need not be measured continuously if it accounts for less than 5% of total emission of nitrogen oxides. The proportion of nitrogen dioxide should then be simply calculated.
[c] This limit value applies to all fluidised bed plants irrespective of thermal capacity.
[d] For fluidised bed combustion plants the Commission will make appropriate proposals later.
[†] In June 1988, the EEC Council of Environment Ministers agreed a Directive to limit emissions of SO_2, NO_x and particulate from plant above 50 MW. It includes specific emission limits for new plant, and national limits for emissions from existing plant.

Table 5.2.3 EEC proposed directive on large combustion plant

Detail reservation	General reservation
Belgium	Greece
Denmark	Ireland
France	Italy
Federal Republic of Germany	*United Kingdom*
Luxembourg	
Netherlands	
	New members
	? Portugal
	? Spain

annual sulphur dioxide emissions, compared to 1980. They failed in this endeavour, which they stated to be all that Spain could offer in this respect.

It has become an important objective of the Community to make progress on the Directive, and several sets of modifications have been suggested to the original proposals under successive presidencies, including that of the United Kingdom. By mid-1987, variations had been proposed to the reductions that would need to be borne by individual member countries in an attempt to make the Directive more politically acceptable. The level of emission reduction required for plant of different size has also been varied in some proposals, with a sliding scale suggested linking plant size and allowable emissions. It seems possible that the smallest plant to be included in the Directive could be increased from 50 MW to 100 MW. The size at which the maximum emission reduction is required might similarly be increased above 300 MW.

More fundamentally, some have questioned whether the Directive should be split into two, respectively addressing overall reductions of emissions and plant emission limits. Despite this movement, and periodic bursts of optimism when some variant of the Directive has seemed near to agreement, there must be doubts about when a decisive break in the political logjam will occur. What is clear, particularly as membership of the Community has expanded, is that reservations by the UK are just one factor among several in explaining the lack of progress.

5.3 PROSPECTS FOR EMISSIONS OF ACID RAIN PRECURSORS IN THE UK

5.3.1 Introduction

The purpose of this paper is to indicate the ways in which emissions of the main precursors of acid rain

might evolve in the next 25 years. The substances considered are sulphur dioxide, nitrogen oxides and hydrocarbons (HC). The findings presented relate to 'unconstrained' emissions, those which would result if no new environmental laws or regulations are passed. Besides defining the possible ranges of future atmospheric loadings which would result if no legislative action were taken, this approach enables us to:

(i) define the possible starting points for any active emission abatement measures; and
(ii) identify the amount of abatement required to comply with specific emissions reduction proposals, such as those contained in the EEC draft directive on emissions from large combustion plant or those implied by membership of the 30% Club.

5.3.2 Uncertainty in emission levels

There will be a considerable degree of uncertainty attached to any set of emissions projections. These are generally obtained by applying a set of appropriate emission factors (which are defined as the amount of pollutant emitted per unit of fuel burned) to a set of energy projections. The uncertainty in future emission levels obviously extends to the amount of abatement required to achieve aggregate emission targets, such as those proposed by the EEC Commission. The uncertainty derives from:

(i) uncertainty about both the current and the future qualities of fuels. This is particularly important when considering SO_2 emissions which are closely related to the sulphur content of fuels;
(ii) lack of accurate knowledge about combustion conditions and hence emission factors. This applies particularly to emissions of NO_x and HC and makes historic as well as future emission levels uncertain. The recent 40% upward revision in NO_x emissions from the transport sector by the Department of the Environment illustrates this point; and
(iii) the uncertainty inherent in the underlying energy projections.

The evidence submitted to the Sizewell 'B' Public Inquiry, the last forum in which many of the energy options for the UK were discussed, shows how wide the range of uncertainty about future energy demand and supply can be. These uncertainties may be divided into two distinct classes.

(1) There is the uncertainty about future levels of world economic growth and energy prices which will be reflected in UK economic activity, energy demand and, hence, emission levels. These factors are effectively beyond the control of a national government.

(2) There is, however, the uncertainty about which policies will be adopted to influence energy demand and ensure that demand is met. This uncertainty is about decisions which remain to be taken by this and future governments and the energy supply industries. Examples of the decisions which will affect future emission levels include those on the depletion of North Sea oil and gas resources, the future of the coal industry, levels of coal imports, energy conservation policy and the degree of reliance on the nuclear option for electricity supply.

The uncertainty about future energy demand and supply is addressed explicitly in this paper, and the other sources of uncertainty, which are less significant, are referred to where appropriate.

5.3.3 Sources of emission projections

The starting point for the emission projections has been the evidence which the Department of Energy (DEn) submitted to the Sizewell Inquiry. However, these projections have been used as a starting point only, as both circumstances and perceptions have changed in the last four years. This was the case even before the significant developments of 1986, the Chernobyl nuclear accident and the major fall in world oil prices, which highlighted the wide range of views held about the future development of nuclear power and raised the possibility of oil returning to play a larger role in meeting energy needs. The principal modifications made to the DEn projections are:

(i) a greater use of natural gas in the industrial and commercial sectors;
(ii) slight adjustments to future levels of energy demand to take account of the recession which dominated the early 1980s (note, however, that the long run rates of growth were not adjusted);
(iii) a five-year extension of the operating lives of Magnox nuclear reactors; and
(iv) a slower rate of ordering for Pressurised Water Reactors (PWRs).

The modified DEn projections have been turned into emission figures using a methodology identical

to that used by the Warren Spring Laboratory to derive historical emission estimates published by the DoE. For NO_x and HC, Warren Spring emission factors have been used while, for SO_2, the petroleum, coal and electricity supply industries have been consulted about the possible future qualities of fuel to be burned.

This paper initially considers a conventional view of future energy sector developments and emission levels based on the modified Sizewell projections. The projections are then modified further to consider the possible impacts of energy policies which are rather different from those implicit in the DEn's submission to the Sizewell Inquiry.

5.3.4 Future SO_2 emission levels

First, the ranges in which future emission levels might lie under a conventional set of energy policies are considered. Figure 5.3.1 shows SO_2 emissions back to 1970 and possible future levels as far ahead as 2010, with a band of uncertainty due to energy demand included. (Note that Fig. 5.3.1 does not take account of the CEGB's 1986 proposal to fit flue gas desulphurisation (FGD) equipment to 6000 MW of coal-fired capacity. The consequences of this measure are explored later.)

The 40% reduction in emissions since 1970 is clearly shown. Three quarters of the past fall in emissions comes from manufacturing industry and may be attributed to: (i) industrial restructuring; (ii) energy conservation; (iii) the increased use of sulphur-free natural gas; and (iv) the reduced sulphur content of fuel oils.

Taking the period up to 1990 first, SO_2 emissions are expected to drop, though not as rapidly as they have done since 1979. Further decline should occur because of: (i) continuing energy conservation; and (ii) the fact that 5500 MW of nuclear capacity will

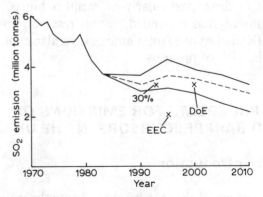

Fig. 5.3.1. Past and projected SO_2 emissions.

have been commissioned during the 1980s, displacing large quantities of coal.

Note that industrial restructuring and the falling average sulphur content of fuel oil, which have been factors in the past decline in SO_2 emissions, are not expected to make major contributions in the future. Some degree of industrial growth has now resumed and this could compensate, at least in part, for the energy conservation which has been, and still is, under way. The reduced sulphur content of fuel oil has been linked to the increasing use of low sulphur North Sea crude oils in UK refineries and to the fact that the higher sulphur fuel oils have been run through refinery cracking plant as fuel oil demand declined in the early 1980s. These reductions have slowed down, and any further reductions could be threatened if demand for fuel oil picks up as a result of lower prices.

The role of natural gas in the UK energy markets, particularly the industrial market, is difficult to predict because of the recent drop in oil prices. Prior to the price fall, the British Gas Corporation had expected to continue to gain market share from oil in the industrial and commercial sectors. This would have tended to push SO_2 emissions downwards. However, there are indications that, in the early part of 1986, some consumers switched from gas to oil, leading to increases in emissions.

In the longer term, the electricity supply industry, which accounted for 70% of 1983 SO_2 emissions, will largely determine the pattern of changes in future emissions. During the early 1990s, when no new nuclear plant is scheduled to come on-line and the first Magnox reactors may be decommissioned, power station coal use is likely to rise. In the absence of any FGD equipment being fitted to existing power stations, this would lead to an increase in emissions. Thereafter, what happens depends on the rate of ordering of new power stations. The central scenario in this conventional look at future emission levels assumes that Sizewell will be built and that 5500 MW of additional plant will be completed by the end of the century. In this case, emissions peak about 1995 and then begin to fall off quite rapidly. The question of new power plant ordering and emission levels is addressed below.

Figure 5.3.1 also shows how future SO_2 emission levels might compare with the 30% Club target of a 30% emissions reduction between 1980 and 1993 and the DoE's 'policy aim' of a 30% reduction below the 1980 level by the late 1990s. Both targets lie within the band of uncertainty in Fig. 5.3.1.

However, only an adverse combination of circumstances — low economic growth and high world energy prices — could lead to the targets being met. Assuming a central case, the degree of active emissions abatement required to meet the 30% Club target by 1993 would be slightly larger than would be the case for the DoE's later target date because unconstrained emissions could still be rising after 1993.

There is no possibility whatsoever of meeting the EEC's proposed 60% reduction in SO_2 emissions from large combustion plant by 1995. (Large combustion plant is defined as that having a thermal output of more than 50 MW and would include all power stations, most refineries and a quarter to a third of fuel use in manufacturing industry.) The 1995 target shown in Fig. 5.3.1 allows for the split between large and small combustion plant. Achieving the 1995 target would require the fitting of FGD equipment on eight to twenty power stations, depending on exactly how unconstrained emissions evolved. Assuming that a major PWR programme is undertaken, postponement of the EEC compliance date would reduce the amount of active SO_2 abatement required.

Figure 5.3.2 shows, in index form, how SO_2 emissions from power stations, manufacturing industry (including refineries) and other sources would evolve in one of the central scenarios from the DEn's Sizewell projections. (The scenario taken is 'BL', incorporating 1·5% GDP growth with a major contribution from the service sector and the lower world oil price background.) By 1983, industrial emissions had fallen 40% below the 1980 level. Although an increase could reasonably be expected through a resumption of economic growth, in this central case emissions remain at about 70% of the 1980 level, staying in line with the 30% Club target. The reduction in miscellaneous emissions (from households, transport, etc.) has been less than for

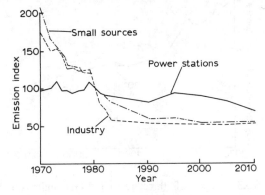

Fig. 5.3.2. SO_2 emission indices (1980 = 100).

industry, but they too may be expected to be about 30% below 1980 levels by the early 1990s. However, unconstrained power station emissions are unlikely to have fallen to this extent until well into the next century.

In September 1986, the Government authorised the CEGB to fit FGD to 6000 MW of existing coal-fired power stations. The authorisation followed a recommendation from the CEGB, 'to ensure that taking one year with another SO_2 emissions will steadily fall between now and the end of the century.'

At the same time, the Government announced that FGD would be required for all new coal-fired power stations. The effect of this decision on SO_2 emissions in the central scenario is shown in Fig. 5.3.3. The timing of the three proposed FGD units (one coming on-line in 1993, one in 1995 and one in 1997) would not quite allow the 30% Club target to be met but would secure the DoE's policy aim of a 30% reduction by the end of the 1990s. The UK would still be some way from the EEC Commission proposal for a 60% reduction in emissions from large combustion plant.

5.3.5 NO$_x$ emissions

Figure 5.3.4 shows past NO_x emissions and possible future emissions under the same range of economic assumptions. The recent reductions in NO_x emissions have been less than for SO_2, and whether they stay roughly level or decline further depends on future levels of economic growth. However, a central view would be that NO_x emissions could fall by 1990 and remain steady until about 1995. Thereafter, significant declines could occur because of a reduced coal-burn by the electricity supply industry if nuclear capacity is expanded. The application of tighter EEC regulations on emissions from new

motor vehicles will push NO_x emissions from the transport sector downwards throughout the projection period.

Interestingly, Fig. 5.3.4 shows that the DoE's policy aim of a 30% reduction in total NO_x emissions is slightly more stringent than the EEC proposal for a 40% reduction from large combustion plant only. This follows from the high contribution of transport to total emissions (40% in 1983). Despite this, the DoE's aim lies just inside the range of possible emission levels while the EEC proposal is outside because of the timing of the targets. However, on a central view of emissions, active abatement measures would be required to meet both targets. It is only in the transport sector that a 30% reduction in NO_x emissions by the late 1990s might be achieved. This would not be the case for either the power station or industrial sectors, from which significant emission reductions would be required to meet the EEC 40% reduction target.

5.3.6 HC emissions

Current HC emissions are extremely uncertain. Figure 5.3.5 shows estimated emissions assuming that about 1% of natural gas production is emitted because of leakage. This figure is the result of recent

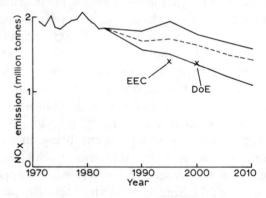

Fig. 5.3.4. Past and projected NO_x emissions.

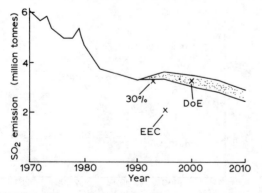

Fig. 5.3.3. Effect of three retrofit FGD units on SO_2 emissions.

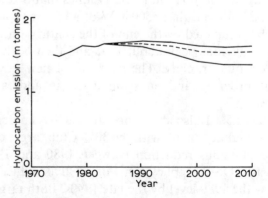

Fig. 5.3.5. Past and projected hydrocarbon emissions.

discussions between the Warren Spring Laboratory and British Gas. In 1983, it is estimated that 36% of HC emissions resulted from industrial processes (unrelated to energy use), 33% from road transport and 23% from gas leakage. Power stations and industrial energy use are thus fairly insignificant sources of HC. On a central view of future economic developments, emissions would be expected to remain fairly flat in the near future, but decline slightly during the 1990s. EEC regulations on vehicle emissions should reduce HC emissions from the transport sector. A further discussion of hydrocarbon emissions follows the section on vehicle emissions.

5.3.7 The impact of alternative energy policies

So far, a conventional view has been taken of developments in the energy sector. However, the choice of different strategies for both meeting and influencing energy demand has significant implications for unconstrained emission levels. The main alternatives which are discussed here are the role of nuclear power, the greater use of renewable energy resources and a 'low energy' future incorporating a greater degree of reliance on energy conservation. These topics have been the topic of previous Watt Committee reports.

The above alternatives would affect emission levels in a major way. There are other alternatives, important in themselves, which would have a smaller influence on emission levels, generally pushing them down. For instance, the use of 5 million tonnes a year of low-sulphur imported coal at a base load power station could cut SO_2 emissions by 70 000–80 000 tonnes annually. In addition, full use of the new 2000 MW Channel link with France would cut SO_2 emissions by around 150 000 tonnes and NO_x emissions by 40 000–50 000 tonnes. The British Gas Corporation has recently indicated that it could increase industrial gas sales by 800 million therms year^{-1} (indications given, however, just before the oil price fell and squeezed the industrial gas market). This could, depending on which fuels were displaced, cut SO_2 emissions by 70 000–90 000 tonnes and NO_x emissions by 20 000–30 000 tonnes. These figures may be compared with the 130 000–140 000 tonnes of SO_2 abatement which would be obtained by installing FGD equipment on a single base load coal-fired power station and the 20 000 tonnes of NO_x abatement which could be achieved by installing low-NO_x burners.

The first major energy policy alternative examined concerns the rate of ordering of nuclear power plant. It is unlikely that a more rapid rate of nuclear power plant construction than that assumed already would be possible until the next century. Therefore, Fig. 5.3.6 shows that an increase in the ordering rate for new nuclear plant could not have a major impact on SO_2 emission levels until after the year 2000, too late to help meet the 30% Club target or the EEC Commission proposals. A halt to nuclear plant ordering (including the cancellation of Sizewell) would lead to only small increases in emission levels after 1990. This is because the alternative to nuclear power is coal-fired plant fitted with FGD which could remove more than 90% of the SO_2 in flue gases. However, a rapid phase-out of existing nuclear power stations would cause emissions to rise more significantly.

Greater use of renewable energy resources, such as wind or wave power, would make a contribution to keeping emissions down. However, the potential contribution of renewables within the EEC time-scale for action on reducing emissions is minimal. All the renewable sources (apart from a Severn Barrage scheme) are on a small scale and would require the availability of backup (fossil) generating capacity to allow for low output during unfavourable weather conditions. At best, the renewable energy sources could have a small contribution to make at individual locations where conditions are particularly favourable for their use.

On the other hand, following a 'low energy' strategy would begin to have an impact on emission levels from an earlier date. In the sense that there is no direct one-to-one relationship assumed between energy use and economic activity, the DEn's Sizewell projections incorporate some degree of energy conservation which reflects measures stimulated by changes in the real cost of energy. However, the establishment of the Energy Ef-

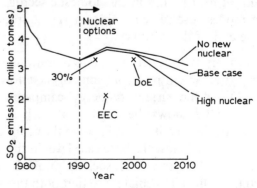

Fig. 5.3.6. SO_2 emissions and nuclear power.

ficiency Office (EEO), the designation of 1986 as 'Energy Efficiency Year' and the current 'monergy' campaign indicate that more energy conservation is considered possible than that which would be induced by price incentives alone.

Energy conservation covers a wide range of technological options which would displace different types of energy use and would take effect over different time-scales. For example, loft insulation is a measure which could be implemented in the short term and which would reduce the use of gas and electricity. On the other hand, the greater use of combined heat and power (CHP), as exemplified by the trial schemes in Leeds, Edinburgh, Belfast and Newcastle, is a measure which would have an impact in the longer term, would also reduce the domestic use of gas and electricity but would increase the use of coal at the CHP stations.

For these reasons, any energy conservation programme would necessarily be a complex package of individual measures. This makes it extremely difficult to quantify the potential effects of energy conservation on emissions. One of the few examples of work carried out to assess the impacts of a large-scale energy conservation programme is Gerald Leach's study of low energy futures (Leach, 1979). Particularly in the industrial sector, many of his conclusions about energy conservation potential now appear conservative. Nevertheless, for the domestic, commercial and institutional buildings sectors, Leach's scenarios imply the use of considerably less energy than do those of the DEn for a given level of economic activity.

The main measures which Leach assumes are as follows:

(i) tightening building regulations to improve thermal performance of public buildings;
(ii) energy performance standards for cars and domestic appliances;
(iii) greater use of CHP and district heating;
(iv) use of heat pumps in the domestic sector;
(v) a modest use of renewable energy sources by the end of the century.

This strategy is considerably more ambitious than current government policy, in spite of the efforts of the EEO and the current 'monergy' campaign.

Figure 5.3.7 shows the impacts of such a programme on SO_2 emissions. This would result in a 30% reduction below the base case level in the year 2000. Some of the measures, such as loft insulation, would take effect immediately so that both the 30% Club target and the DoE's policy aim of a 30%

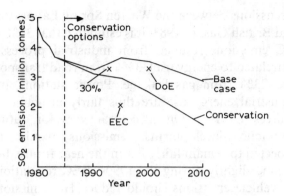

Fig. 5.3.7. SO_2 emissions and energy conservation.

reduction by the end of the 1990s would be met. Once the longer term measures work through the system, SO_2 emissions some 40% below those following from a more conventional view of the future are suggested. However, it would still not be possible to achieve the EEC's proposed target of a 60% reduction in SO_2 emissions from large plant by 1995, though it might be possible to do so by the first decade of the next century.

This discussion of the impacts of energy conservation on emission levels is necessarily brief, but it does indicate the potential that exists.

5.3.8 Conclusions

To summarise, in spite of the many uncertainties, the following conclusions can be reached:

(i) the long-term trend in SO_2 emissions is downwards. The recent decision to retrofit 6000 MW of existing coal-fired power station capacity with FGD will prevent a temporary increase in emissions in the mid-1990s when there is a pause in the commissioning of new base load power plant;

(ii) NO_x emissions should also drop, partly through tighter vehicle emission standards, though again there may be a pause in the early 1990s;

(iii) HC emissions are likely to be fairly flat, but vehicle emissions should fall, again because of EEC regulations;

(iv) the 30% Club target will be attained only under adverse economic circumstances, if a vigorous energy conservation programme is initiated or if a combination of other measures, such as importing low sulphur coal, using more natural gas or building a second Channel electricity link, is undertaken.

(v) The FGD retrofit decision virtually guarantees

the achievement of the DoE's policy aim of a 30% reduction in SO_2 emissions by the end of the 1990s.

(vi) the emissions reductions proposed by the EEC for SO_2 and NO_x could not be obtained without active abatement measures under any foreseeable combination of circumstances;

(vii) in the longer term, beyond the end of the century, there is a clear prospect of significantly reduced levels of all types of emissions.

REFERENCE

LEACH, G. (1979). *A Low Energy Strategy for the UK*. The International Institute for Environment and Development, London.

5.4 TECHNOLOGIES FOR CONTROLLING ACID GAS EMISSIONS

5.4.1 Introduction

This paper provides an up-date on material in Watt Committee report No. 14[1] on technologies and strategies for emission reduction from stationary sources. It assumes a familiarity with the discussion on these subjects in that report. The paper is mainly concerned with technologies for treating flue gases from combustion, prior to their release to the atmosphere, although other approaches are reviewed briefly. There is some discussion of developments in sulphur dioxide emission control and particular consideration of byproduct management problems which arise from this. The main focus of the section is, however, on a subject not covered in much detail in the previous report — the control of emissions of nitrogen oxides. The paper primarily concerns itself with technologies appropriate for application at large-scale facilities ($\geq 300\,MW(e)$), although there is some discussion of smaller scale strategies. It is also primarily addressed to applications for coal-fired facilities, since in the UK context it is these which dominate the debate on emissions from large-scale stationary sources.

In response to impending legislation in several countries, intended to curtail severely emissions of sulphur dioxide and nitrogen oxides, especially for new plant, there has been considerable technical progress since Report No. 14 was published.[2] Consequently there are numerous references in this paper to experience in overseas countries, especially in Europe. The varying range of circumstances in these countries means that a variety of technologies and strategies is being deployed to provide answers

to the particular situations which different national regulations, different fuels and different local characteristics create. For this reason, care must be exercised in translating evidence from operational experiences in one country to those in another. The discussion of overseas practices in this section should not be taken to imply that these necessarily can, or should, be directly applicable to the UK.

Putting aside cost considerations, it is now possible, *technically*, to design new fossil fuel plant with very low emission levels of acid gases. However, there are unavoidable costs involved in achieving such levels and well-established principles of escalating marginal costs operate for each incremental reduction in emission levels that may be required. These relate not only to higher capital and operating costs but also to possible reduced overall availability of plant fitted with emission control equipment whose long-term reliability has still to be demonstrated. For these reasons, the levels of emission reductions imposed by different regulatory requirements will have a significant impact on the overall costs of meeting those requirements. These regulations are discussed elsewhere in this report.

5.4.2 Fuel selection and energy conservation as emission control strategies

As the Watt Committee Report No. 14 noted,[1] the use of emission control technologies is but one of a number of options which may be components of a strategy for reducing acid gas emissions from combustion sources. Switching to use of less polluting fuels or the reduction of energy requirements through energy conservation programmes can also be adopted to control overall pollutant production levels. However, to date these strategies have not been embarked upon to any major degree. The reasons vary from country to country but, in the case of fuel switching, often relate to economic and social disruptions which would follow large-scale changes in the fuel base of countries' energy economies, to fears of making a country's energy economy vulnerable to externally imposed supply difficulties or to environmental uncertainties associated with possible substitutions. For a discussion of this debate in the USA, see Ref. 3.

Furthermore, the unpredictable nature of energy markets over the past few years has created a sense of uncertainty about the best economic options for the future. This, coupled with economic recession, has led to plant operators adopting a cautious

Table 5.4.1 Environmental consequences of fuel substitutions

Substitute fuel	Existing fuel				
	Average (around 1·5%) S coal	Low (under 1%) S coal	High (over 2·5%) S oil	Low (under 1%) S oil	Natural gas
Average S coal	1.[a] Possible in short term to exploit minor differences in fuel characteristics 2. Easy 3. May be done to exploit higher ash retention of S and better fuel N emission characteristics	1.[a] Unlikely	1.[a] Possible in medium term 2. Significant modifications required to boiler—downrating, improved particulate control, new firing system 3. Lower sulphur dioxide but increased particulate emissions. Possible lowering of nitrogen oxides if coal supplied as coal/water mixture	1.[a] Likely only if substitute fuel considerably cheaper than existing fuel 2. Significant modifications required to boiler—downrating, improved particulate control, new firing system 3. Increased formation of all pollutants	1.[a] Likely only if substitute fuel considerably cheaper than existing fuel or as gas conservation policy 2. Significant modifications required to boiler—downrating, improved particulate control, new firing system 3. Increased formation of all pollutants
Low S coal	1. Possible in short term 2. Minor modifications may be necessary, e.g. upgrading of particulate control 3. Sulphur dioxide emissions reduced, nitrogen oxides emissions probably unchanged	1. Possible in short term to exploit minor differences in fuel characteristics 2. Easy 3. May be done to exploit higher ash retention of S and better fuel N emission characteristics	1. Possible in medium term 2. Significant modifications required to boiler—downrating, improved particulate control, new firing system 3. Lower sulphur dioxide but increased particulate emissions. Possible lowering of nitrogen oxides if coal supplied as coal/water mixture	1. Possible in medium term 2. Significant modifications required to boiler—downrating, improved particulate control, new firing system 3. Lower sulphur dioxide but increased particulate emissions. Possible lowering of nitrogen oxides if coal supplied as coal/water mixture	1. Likely only if substitute fuel considerably cheaper than existing fuel or as gas conservation policy 2. Significant modifications required to boiler—downrating, improved particulate control, new firing system 3. Increased formation of all pollutants
Low S oil	1. Unlikely	1. Unlikely	1. Possible in short term 2. Minor modifications necessary 3. Reduced sulphur dioxide emissions	1. Unlikely	1. Unlikely
Natural gas	1. Possible in medium term. Opportunities currently restricted by regulations against gas burning in several countries 2. Minor modifications probably necessary 3. Lower sulphur dioxide and particulate emissions; also nitrogen oxides in smaller scale plant.	1. Possible in medium term. Opportunities currently restricted by regulations against gas burning in several countries. Probably less likely than substitution for higher S coal 2. Minor modifications probably necessary 3. Lower sulphur dioxide and particulate emissions; also nitrogen oxides in smaller scale plant.	1. Possible in short term 2. Minor modifications 3. Lower sulphur dioxide and particulate emissions	1. Possible in short term 2. Minor modifications 3. Lower sulphur dioxide and particulate emissions	1. Unlikely

[a] 1. indicates likelihood of fuel substitution, 2. ease of substitution, 3. issues/consequences for emission.
Source: from compilation by P. W. Dacey

approach to changing the energy basis of their operations. However, it is possible to identify the likely direction of certain fuel switching changes and the consequences which would arise from substitution of one fuel for another. These are detailed in Table 5.4.1, which considers possible substitutions for control of emissions.

With energy management programmes, one factor is the comparatively long time-scale which such strategies require to have significant reduction impacts. Several countries, including the UK,[4] have observed that increased energy efficiency has contributed to emission reductions achieved in the past decade or so but until recently there has been no *conscious* attempt to pursue energy conservation because of its *environmental*, as opposed to straightforward economic, consequences. However, some countries have now begun to incorporate environmental targets into energy management programmes, with one of the most prominent efforts being made in Denmark.[5] This country is highly dependent on external sources of energy while at the same time sharing the common Scandinavian concern about the environment. The emphasis on such a strategy, incorporating in particular fairly rigid plans for provision of space heating, is therefore understandable.

The contribution that energy management can make to emissions reductions is highly dependent on factors such as the energy use patterns of individual countries, the stock of plants which provide the energy, the emission characteristics associated with the saved energy, location of the saved energy (where local environmental considerations are significant) and even the time of day at which the energy is saved, in situations where a range of fuels is used to provide different tranches of electricity. However, as a general rule, the continuous operation of plant at optimum efficiency will minimise fuel use and pollutant emissions, so that strategies which reduce energy losses between the primary supply and final use and the need to start up and shut down plant frequently, will be potentially attractive.

5.4.3 Control of sulphur dioxide emissions

5.4.3.1 Flue gas desulphurisation

In the area of control of sulphur dioxide emissions, since the Watt Committee Report No. 14 was published, there has been rapid deployment in Europe, particularly in West Germany, Austria and to a lesser extent, Sweden and The Netherlands, of 'second generation' flue gas desulphurisation (FGD) systems, mainly those using lime or limestone as the SO_2 reagent, producing gypsum $(CaSO_4.2H_2O)$ as a byproduct material. Earlier designs of FGD, installed in the USA, which produce an unattractive, essentially unusable, calcium sulphite sludge $(CaSO_3.1/2H_2O)$, have not found favour in Europe. (Only one such first generation plant has been installed in Europe, in West Germany, and this has subsequently been converted to a gypsum-producing system.) For this reason, the CEGB has been assessing limestone-gypsum systems for application in the UK, extending its interest beyond regenerable systems alone, as stated in Watt Committee Report No. 14. Some more recent data on plant characteristics and costs are given in Table 5.4.2. It has been indicated that these data have been superseded as the CEGB moves towards actual ordering of FGD; however, it should continue to provide a reasonable guide to the order of costs of the processes under consideration.

Sole reliance on regenerable systems producing sulphur or sulphuric acid as a saleable byproduct could cause market saturation problems. However, limestone/gypsum systems are also not without their byproduct management problems and because of the increased number of these systems in Europe, there has been a marked growth in interest in possible secondary pollution issues associated with byproduct disposal from FGD plants, despite the fact that the intention has frequently existed, and has actually been implemented in a number of countries, to use the waste gypsum produced in various applications such as additives to cement or for wall board manufacture. In the UK, the CEGB has indicated an intention to produce wallboard quality gypsum and to seek market outlets for its products. The question of input and output product management is discussed in more detail below.

In the other main category of technologies for FGD —regenerable systems, producing either concentrated streams of sulphur dioxide, sulphuric acid or liquid sulphur, for subsequent re-use—orders have been placed in three countries in Europe for the British Wellman–Lord system, producing sulphur dioxide and sulphur. In West Germany, the first plant using a novel FGD system based upon activated carbon—the Bergbauforschung–Uhde system—is under construction. This technology also has the ability to reduce emissions of nitrogen oxides, about which more is written below. This system might be compared with a giant gas mask or

Table 5.4.2 Costs of flue gas desulphurisation at a 2GW power station

Process type	Plant details[a]	Major product and amounts (tonnes)	Capital cost (£ million)	Net annual operating & maintenance (£ million) costs[b]
Regenerable	New Wellman–Lord system on 3 × 660 MW(e)	Sulphur 68 000	123	4·6[1]
Regenerable	Retrofitted Wellman–Lord system on 4 × 500 MW(e)	Sulphuric acid at 96% concentration 230 000	160	1·6[2]
Limestone/ gypsum	Unspecified on 3 × 660 MW(e)[c]	Gypsum 520 000	123	7·3[3]

[a] Station parameters are 2%S, 0·4% Cl coal, 70% load factor. These are conservative assumptions.
[b] In each case, the FGD is assumed to be operating at 90% removal of SO_2. All costs are at 1983 levels.
1. Operating/maintenance costs are high due to energy costs of producing elemental sulphur. Byproduct sales credit is assumed.
2. Operating/maintenance costs are lower than first case because sulphuric acid produced. Byproduct sales credit is assumed.
3. Operating/maintenance costs are high because no sale of byproduct assumed. Byproduct is disposed to landfill at zero cost.
[c] Reference source does not identify whether these data refer to new or retrofit station.
These data may also be compared with more detailed individual station information given in Ref. 6. Estimates of cost as presented here may be revised as plans for FGD are progressed within the CEGB.
Source: Reference 10.

cooker hood and uses the ability of activated carbon to adsorb large quantities of those gases.

There have also been the first orders and construction works of spray-drying type FGD systems in Germany, Austria and Denmark, both for large- and smaller-scale applications. These FGD systems use a concentrated slurry of slaked lime to absorb sulphur dioxide—the water present is completely evaporated by the heat of the flue gas to give a dry powdery substance, closely resembling dried milk (which is also frequently produced using spray-drying technology). The reacted sorbent is predominantly calcium sulphite ($CaSO_3.1/2H_2O$), a material with limited re-use applications and possible disposal problems. In all cases to date the plants are disposing of their wastes to convenient land-fills; the limited prospects of using wastes from spray drying systems remain a disincentive against a wider deployment.

In the UK, the CEGB has recently announced that it is to fit all new coal-fired power plant (of which three 1800 MW(e) stations had been announced by mid-1987) with FGD systems and that, additionally, three existing 2000 MW(e) stations will be retrofitted with sulphur dioxide removal equipment. Evaluation of limestone/gypsum and regenerative systems is currently proceeding, based on engineering evaluations, economic assessment and environmental considerations. The plant will be required to operate at around 90% sulphur dioxide removal efficiency.

As well as sulphur dioxide and nitrogen oxides, untreated flue gases contain amounts of hydrochloric acid gas, in quantities related to the chloride content of the coal burned. UK coals have a chloride content higher than many overseas coals, averaging around 0·25%, but levels up to 0·8% occur. This must be removed before the flue gases enter a regenerable type of FGD, otherwise the chlorides build up in the reagent cycle, causing unacceptable loss of efficiency. For once-through limestone/gypsum systems, the presence of chlorides is not a major operational problem but high levels in the end-product may limit its marketability.

For these reasons, it is generally desirable to remove the hydrogen chloride from the flue gas before the larger gas treatment for sulphur dioxide. For high chloride coals, this could lead to large quantities of effluent. The CEGB has been working on a process which produces a smaller volume of concentrated calcium chloride solution. This is being tested on a 5 MW(e) pilot plant constructed at Ratcliffe on Soar power station in Nottinghamshire.

5.4.3.2 Other emission control technologies

Turning to technologies for sulphur dioxide control

other than FGD, there has again been a number of developments since the Watt Committee Report No. 14 was published. As the report noted, it is possible to achieve some reductions in sulphur dioxide formation by directly injecting pulverised limestone, along with—or in the vicinity of—fuel, into pulverised fuel furnaces. In both Germany and Austria, demonstration of direct injection of pulverised limestone into furnaces for sulphur dioxide control has achieved reasonable levels of reduction (50% or thereabouts) but with associated problems of waste disposal, particularly the alkaline nature of leachates from disposal sites. The reacted sorbent is collected with associated fly ash in existing or upgraded particulate collection systems. It is essentially unutilisable and must be deposited in landfills.

The best sulphur dioxide removal levels achieved with this technology are considerably below those likely to be standard requirements for large-scale plant and the technology is less economic in its use of limestone (typically three times less so) than FGD. Waste disposal problems have already been discussed. Not surprisingly, in Europe, the furnace sorbent injection systems are being superseded by retrofitting of FGD systems.

In non-pulverised fuel systems (i.e. for smaller scale applications), a lot of work is being carried out on various forms of introducing limestone as a sulphur dioxide sorbent in the combustion bed. With conventional (e.g. stoker-fired) systems, some work has been done but the levels of sulphur dioxide reduction achieved by addition of limestone to the bed or injecting it over the bed have not been very significant.

Sulphur dioxide emissions from fluidised bed plant are reduced by the presence of an in-bed sorbent, usually limestone or dolomite (calcium magnesium carbonate). Unlike with conventional FGD plant, a separate sulphur dioxide removal stage is thereby avoided. The disadvantage is that sorbent utilisation is less efficient than in FGD systems—90% removal of sulphur dioxide requires about twice the quantity of sorbent. As a result, a larger volume of solid wastes is produced—a mixture of ordinary coal ash, unreacted sorbent and reacted sorbent in the form of calcium sulphate anhydrite. This material is produced dry but will probably require landfilling in a manner similar to conventional coal ash.

In Europe, sulphur dioxide reductions of 85% or more are being achieved in commercially deployed fluidised bed plants. Attention is now turning to scaling these plant up to larger size applications but the technology is likely to be limited to the small- to medium-scale of application.[6]

5.4.3.3 Advanced technologies

In the course of the Parliamentary inquiries into 'acid rain' which have taken place during the past four years, there has been considerable discussion of completely new types of power generation facilities which are said to hold out the attractive option of low emissions *and* improved efficiencies of converting heat to electrical energy.[7,8,9] In the UK, these discussions have tended to be dominated by consideration of pressurised fluidised bed combustion combined cycle (PFBC CC), probably because of the location of the pressurised fluidised bed test facility at Grimethorpe mentioned in the previous Watt Committee report. Originally developed under International Energy Agency auspices, the second stage of the demonstration project of that plant is now going ahead with joint funding in excess of £27 000 000 from the CEGB and British Coal (formerly the National Coal Board). Some specialist work on feeding the combustor with coal/water slurry is also being supported by US sources. The second stage is designed to test modifications to tackle some of the operational problems experienced during the first stage of the project and also to focus on emission characteristics. It is important to realise that this project has not yet progressed to the critical stage of demonstrating the feasibility of linking an exhaust gas turbogenerator to the system, without which the expected conversion efficiencies from heat to electrical energy will not be achieved.

Pressurised bubbling bed technology, of the type under development at Grimethorpe, is only one of several fluidised bed technologies which can be operated in combined cycle, with a gas turbine as well as the usual steam turbine. A number of alternative options is being examined by a steering committee set up by the UK Advisory Council on Research and Development (ACORD), involving the CEGB, British Coal and the Department of Energy. Design studies for a power station based on the pressurised bubbling bed, a pressurised circulating bed and various types of atmospheric circulating bed with air heater and hot air turbine have been subject to technical and economic evaluation. The findings have yet to be published.

Combined cycle technologies certainly give a thermodynamic efficiency advantage over ordinary pulverised fuel combustion plant, especially when these are fitted with FGD. However, even if the technical feasibility of combined cycle is demonstrated, the optimum size of facility remains to be established, since capital costs and complexity of the plant differ considerably from conventional combustion technologies.

Some foreign companies have also been developing PFBC technology, most notably the Swedish company ASEA PFBC, which has been developing a different design of PFBC CC from the Grimethorpe facility, in cooperation with a US utility and equipment manufacturer. Two power stations, in the USA and in Stockholm, Sweden, are being equipped with this technology at the 80 MW(e) scale involving retrofitting into existing plant, and an order has been placed for a plant in Spain. These will be the first demonstrations of the full bubbling bed combined cycle technology with a turbine driven by combustion gases.

In the field of the other main low emission power generation technology—integrated coal gasification with combined cycle (IGCC), the most interesting development in the recent past has been the coming into operation of the Cool Water (the name of the place, not the technology) plant in California. In gasification combined cycle, coal is gasified with a mixture of air or oxygen and steam in a pressure vessel, and the resulting gas is fired in a gas turbine. Heat from the turbine exhaust is used to raise steam in the second stage of the combined cycle.

Sulphur is mainly given off in the gasification process as hydrogen sulphide rather than sulphur dioxide and in a more concentrated form than in combustion gases. It can therefore be readily removed, to efficiencies of over 95% by use of conventional industrial technologies as widely used in the petrochemical and ore-smelting industries.

Development of the Cool Water plant has focussed on emission optimisation from the beginning and has been rewarded with extremely low levels of both sulphur dioxide and nitrogen oxides emissions, comfortably under the severest limits postulated anywhere in the world for medium- to large-scale plants. Because of the thrust of the development programme, it is currently uncertain whether this technology can achieve the reliability and conversion efficiencies required for commercial plant. In particular, to raise the conversion efficiency of gasification combined cycle technologies, gas turbines with firing temperatures higher than those of current industrial gas turbines will be necessary. Several turbine manufacturers are working on such advanced technology and there

has been an upswing in the level of utility interest in IGCC for future power generation, notably in the USA and Japan.

Various types of gasification technology are under development for the gasification stage of IGCC. A front runner is the British Gas/Lurgi slagging gasifier, of which a pilot plant is operating at the British Gas research site at Westfield, Scotland. One environmental attraction of this particular gasifier is that it operates at temperatures where the coal ash is fused to produce a glassy, non-leachable frit. An ACORD steering committee is currently reviewing the application of this technology for power generation in the UK.

The development of the hardware of these advanced technologies throughout the world is being limited by research and development funding constraints, so that evaluation has to be primarily based on design studies. It remains unclear what the likelihood and time-scale of progress of both these technologies to commercial applications will be. For this reason, the recently announced round of UK power stations will be based on conventional generation technology. The advanced technologies also have some disadvantages associated with them —notably the complexity of the process in the case of IGCC, and the nature and amounts of solid wastes for disposal with PFBC CC. However, one factor which is becoming clearer is that both these technologies, partly through inherent features and partly through the thrust of development being focussed on the US utility market, are likely to be commercialised at the medium-scale level of application (200–400 MW(e)), rather than large-scale (2000 MW(e)), and might have their first applications in combined heat and power plants rather than simple electricity generation units.

5.4.3.4 The management of byproducts from flue gas desulphurisation systems

The capture of sulphur dioxide from flue gases requires that the gas be locked up in a form which prevents its subsequent return to the atmosphere. As noted above, it may be either chemically bound into a compound or produced by the FGD system in a concentrated form for further chemical processing. The dominant form of system reacts the sulphur dioxide with limestone (calcium carbonate) to form calcium sulphite, which is oxidised *in situ* to give hydrated calcium sulphate (gypsum) ($CaSO_4.2H_2O$). This process is relatively straightforward and the dominance of its use means that early difficulties have essentially been overcome to provide reliably operating systems.

The quantities of reagent and reaction products required are, however, considerable and this has led to a realisation that there will be significant management issues involved in any major programme of flue gas desulphurisation. For example, in West Germany, where a programme involving installation on 37 GW of plant has been adopted, a total of 3·9 million tonnes of flue gas desulphurisation gypsum is expected to be produced by 1990. This compares with a current annual natural production of 4·7 million tonnes, which is expected to decline to some extent. The problem is that the major part of the FGD gypsum will arise in northern parts of the country, a considerable distance from the gypsum plants located in the south and east. Current forecasts suggest that at best only 3·0 million tonnes of the desulphurisation gypsum will be marketable —the remainder, especially from the large brown coal plants in the Bonn/Cologne area, will have to be deposited (Vereinigung des Grosskraftwerks Betrieber, Essen, pers. comm.).

In the UK, the power plants to be retrofitted with FGD systems, and the new stations recently announced, will be among the largest in the world to be so treated. A limestone/gypsum system on a CEGB station would require about 300 000 t of limestone input a year, producing about 470 000 t of gypsum. The throughput of limestone and gypsum would represent an increase of 13–15% in the total (coal and ash) solids throughput of such a station. There is no difficulty in producing gypsum of a quality sufficient for use in plasterboard or plaster manufacture (indeed, a gypsum can be made considerably purer than the natural material) but a single station would produce 10–15% of the UK's total annual demand for gypsum. Gypsum specifications are particularly strict on chloride content, which adversely affects plaster setting qualities; to achieve levels required—below 100 ppm w/w—extensive washing of the FGD gypsum is required, which produces a dilute wastewater stream requiring further treatment before discharge.

Because of market imperfections, it is unlikely that the total production of FGD gypsum in the UK could be utilised. Provision would have to be made for disposal as a contingency to cover the expected lifetime of the station. Market fluctuations, transport disruptions and the inevitable production of some off-specification gypsum will combine to require at least temporary storage facilities. One of the most marked fluctuations is the seasonal demand for gypsum, linked to the cycle of building construction. Gypsum should be able to be disposed of in land-fills but this may require careful control of hydrogeology because of its slight solubility.

Another consideration is the supply of the calcium carbonate reagent. Economics will require this to be as close as possible to the power station, within the constraint of adequate transport links being available. Operational considerations require the limestone to be as pure as possible, in order to reduce the costs of transporting, processing and using inert material, and because presence of sands and clays in the FGD reagent system has adverse effects on its reliability and lifetime.

All the factors discussed above are matters of trade-offs and are summarised in Table 5.4.3. It is clear that management of these factors will be very dependent on the individual circumstances of each site — for example, its location in relation to reagent supply, market for product and disposal possibilities.

One immediate option in this assessment of trade-offs is to adopt a regenerable FGD system, thereby avoiding the problems of reagent supply and producing a byproduct which should have considerably more attractive utilisation prospects. The major factors in this trade-off are the increased capital cost of such systems and their total reliance for achieving acceptable operating costs on a guaranteed market for their product. CEGB figures presented to the House of Lords Select Committee on the European Communities[7] generally show that costs per tonne of sulphur abated are higher for regenerable systems, but the data do not give details of assumptions on prices for marketable products. It is considerably cheaper to operate regenerable systems producing sulphuric acid rather than sulphur because of the additional energy costs required to reduce the sulphur dioxide. This should be partly offset by the higher price per tonne which elemental sulphur would command in markets, but the choice between the two approaches will be critically dependent on the availability of a local outlet for sulphuric acid.

The CEGB has indicated that it would resolve these management problems in any large-scale programme of FGD installation by a mixture of gypsum and regenerable systems, with some sale and some dumping of gypsum, and the regenerable systems producing a mix of sulphur and sulphuric acid. The actual outcome would depend strongly on local circumstances but it is likely that, as in West Germany, some form of strategic plan, showing

Table 5.4.3 Issues in the management of FGD sorbents and byproducts

Issue	'Once-through' systems	Regenerable systems

Sulphur dioxide sorbent supply	Cost of sorbent—
	   * Quality required
	   * Source in relation to power station site
	 Transport mode hazards
		Cost of initial reagent charge

On–site sorbent management	Bulk storage required Normal bulk handling considerations Minor operating hazards (especially with lime-based systems)	Reagent depletion rate and make-up costs Normal operational considerations of a chemical plant Risks of loss of chemical sorbents (COD, odour) Risks of escape of concentrated sulphur dioxide or sulphuric acid streams
Prescrubber requirement	Depends on whether byproduct intended for utilisation or disposal and whether washing of byproduct is an option	Almost certainly necessary to prevent unacceptable loss of chemical sorbents

Byproduct management	Byproduct produced with intention of disposal or utilisation
	 Byproduct actually disposed or utilised
	 Costs of discharging byproduct from responsibility—
	   * Availability of disposal sites
	   * Transportation costs
	   * Availability of markets
	   * Costs of upgrading byproduct for safe disposal or utilisation
	   * Costs of intermediate storage
	 Transport mode hazards
		On-site storage requirements
		 Market trends for sulphur, sulphuric acid and or sulphur dioxide
		 Transport mode hazards (sulphur dioxide and sulphuric acid)

how the different products will fit into expected markets and analysing any secondary environmental consequences, will be required.

5.4.4 Control of nitrogen oxide emissions

5.4.4.1 Approaches to NOₓ reduction

As indicated earlier, there are essentially four different approaches to the reduction of acidic gaseous emissions including nitrogen oxides. The need for fuels can be reduced in various ways so that less is burnt, fuel which is intrinsically less polluting can be used, pollutants can be prevented from forming during combustion, or they can be removed from the flue gases before they are released to the atmosphere. Various combinations of these approaches are also possible to produce a final emission to atmosphere below specified levels, if individual approaches cannot achieve these.

The role of fuel economies in emission reduction has been discussed earlier in this section and strategies can be adopted which address nitrogen oxides as well as sulphur dioxide.

Nitrogen oxide emissions on combustion arise both from nitrogen in the fuel (fuel nitric oxide) and nitrogen in the air (thermal nitric oxide). Generally only 10–25% of the nitrogen in the fuel is converted to fuel nitric oxide. The nitric oxide emitted is converted to nitrogen dioxide in the atmosphere, and emissions are usually quoted in terms of nitrogen dioxide equivalent. Fuel switching can play only a minor part in nitrogen oxides control. The selection of coals for their inherent nitrogen content forms part of the strategy for nitrogen oxides control adopted in Japan, but the contribution which this makes to overall reductions is minor. British coals have limited variability in their inherent nitrogen content which ranges from 1·3 to 1·9% and averages 1·5%, very close to the figure for sulphur. The processing of fuels to reduce nitrogen oxides emissions is not a viable option. The removal of nitrogen from fuels such as coal, which contain high levels of the element, is to all economic purposes impossible since (unlike sulphur) coal nitrogen is entirely chemically, rather than physically, bound.

Thermal nitric oxide is formed in the heat of the combustion flame and is particularly significant at combustion temperatures about 1450°C, most notably in pf combustion. As a general principle, fuel nitric oxide formation makes a larger contribution than thermal nitric oxide, especially when combustion occurs below the 1450°C level. Unconstrained emissions from large-scale plant tend to be up to 1400 mg $NO_2(Nm)^{-3}$, depending on boiler type, though in individual cases, figures can be more than twice as great (particularly for coal). At smaller scale facilities, emissions are generally low and, because of the changed balance between fuel and thermal nitrogen oxides, the fuel used does make a considerable difference, ranging from natural gas, with emission characteristics around 150 mg$(Nm)^{-3}$ to coal with a figure of around 500 mg$(Nm)^{-3}$. Some designs of stoker plant have emissions as low as 300 mg$(Nm)^{-3}$. Fluidised bed plants can achieve levels below this—figures as low as 200 mg$(Nm)^{-3}$ have been claimed by some European manufacturers of circulating fluidised beds, on account of the inherently low combustion temperatures (around 850°C) and with careful control of air access to fuel (see below).

To control nitric oxide emissions, as with sulphur dioxide, there are options for modifying operations within the furnace itself and for treating flue gases prior to discharge to atmosphere. However, the in-furnace control technologies are relatively more important with nitrogen oxides control than they are with sulphur dioxide. Further treatment of flue gases is generally employed only if very low levels of nitrogen oxides emissions are required, and is relatively much more expensive.

The first category of control involves various forms of combustion modification which reduce the formation of nitric oxide within the furnace itself. These have the attraction that they are relatively cheap to install and to operate. Combustion modifications normally involve no use of reagent, unlike flue gas treatment systems. No direct waste products arise although there may be some indirect consequences on solid particulate characteristics. The overall cost of combustion modifications is about a tenth that of flue gas denitrification systems expressed in terms of tonnes of nitrogen oxides abated. Combustion modifications to large-scale plant work out at well under $10 kW^{-1} (1983) capital costs, unless there are special difficulties. The technology can be retrofitted to existing plant although generally speaking the smaller the plant the higher the cost per kW will be, and the smaller the reduction in emissions. The CEGB has recently estimated the cost of retrofitting low nitrogen oxides combustion burners to all its large (over 1000 MW(e)) power stations, a programme announced in mid-1987, to be of the order of £170 m.

The choice between combustion modification or flue gas denitrification depends on the nature of the regulations under which a particular plant is operating. Japanese regulations which came into effect in 1987, and regulations in the FRG for new plant and even for some of the larger existing plant, all set limits that combustion modifications alone are not sufficient to achieve; additional flue gas cleaning will be required. Current drafts of the EC Directive would also require this for plant commissioned after 1996.

5.4.4.2 Combustion modifications for nitrogen oxides control

The first approach to control of nitrogen oxides formation in pulverised coal fired furnaces is to modify the combustion process by altering the distribution of fuel and air in the burner itself (giving a so-called 'low-nox burner') or possibly in the furnace itself, while maintaining the same overall combustion stoichiometry—the combustion process of carbon and the other combustible substances in the fuel in their reaction with atmospheric oxygen. In particular, combustion modifications tend to reduce the temperature or increase the time over which combustion takes place. As a result of these modifications, or 'staging', volatile nitrogen is released from the fuel in reducing conditions where the nitrogen atoms combine to form N_2 molecules instead of being oxidised to nitrogen oxide. Combustion is completed later in the flame or furnace, where the air/fuel ratio is higher. The overall stoichiometry needs to remain the same, otherwise combustion efficiency will be reduced, leading to adverse operational economics, and undesirable hydrocarbon emissions can result.

The CEGB is collaborating with UK equipment manufacturers in the development of low-nox burners for power station application. Trials on a corner-fired furnace at Fiddler's Ferry power station near Runcorn are nearing completion. Reductions in nitrogen oxides formation of around 40% have been measured. The process attempts to avoid problems from impingement of a reducing flame zone onto boiler tube walls, with the associated increased risk of corrosion, by directing some of the combustion air along the furnace walls.

This particular burner/furnace modification for corner-fired units is only potentially suitable for the third of the large CEGB power stations which are fired in this manner. Others have firing from the face of one wall only, while the largest units at Drax power station have opposed wall firing. Tests of low-nox burners suited to these designs of furnace are also underway. If the tests on these designs of station prove that combustion efficiencies and boiler tube lifetimes are not adversely affected, there will be little reason why low nitrogen oxides burner configurations should not be rapidly introduced in all the large CEGB power stations, thus going some way to fulfil the House of Commons Select Committee on the Environment recommendation that *all* power stations be fitted with low nitrogen oxides burners.[8] This recommendation was not accepted by the government at the time because it was considered that the technology was not then proven in UK conditions.

Nitrogen oxides reduction levels of over 40% may be possible with improved burner designs in new boilers specifically designed for staged combustion. Development work is in progress in the US and Japan as well as in the UK.

An alternative approach to combustion modification involves burning of additional fuel above the main burners in the furnace. In this way, nitrogen oxides formed during combustion of the main part of the fuel are subsequently reduced back to elemental nitrogen by reducing flame conditions higher up the furnace. Work has been carried out using natural gas as the secondary fuel in coal-fired boilers in the USA, a conversion which is very easy

to retrofit, and nitrogen oxides reductions approaching 60% have been reported from Japan using coal as the 'reburning' fuel.[11] As with low-nox burners, fuel 'reburning' will probably achieve the highest levels of nitrogen oxides reduction (around 60%) when used in purpose-designed new furnaces.

The mechanisms of nitrogen oxides production in fluidised bed combustion are less well understood than those in pulverised fuel furnaces but since the combustion temperature is lower (generally less than 900°C) the production of nitrogen oxides is lower. Circulating fluidised beds may also display intrinsic fuel staging. As a consequence, fluidised bed combustion can generally meet lower nitrogen oxides emission limits than conventional pulverised plant with the best combustion controls, although some form of flue gas treatment may still be necessary to meet the strictest nitrogen oxides emission limits.

In IGCC plant, nitrogen oxides are formed by the burning gas in the turbine. However, various techniques have been developed, especially injection of steam into the turbine, to minimise nitrogen oxides emission levels.

5.4.4.3 Flue gas treatment for nitrogen oxides control

There are essentially three classes of flue gas treatment systems for nitrogen oxides. In the first, a reducing agent such as ammonia or urea is injected into the upper, superheater region of a boiler, or into a fluidised bed to react with the nitric oxide, as described below. No catalyst is necessary for this reaction. Trials on this technology are currently occurring in the USA and Japan.

The second class comprises technologies specifically and solely designed to remove nitrogen oxides by catalytic reaction with ammonia downstream of the boiler, and the third those installations which are designed to remove sulphur dioxide and nitrogen oxides simultaneously. This class may be subdivided into those plant which have been originally designed to do this joint removal and plant which were originally FGD units alone but where relatively minor modifications are able to be made to the nature of the plant to give an appreciable level of nitrogen oxides removal as well. In these plant, the level of nitrogen oxides removal is seldom as high as that achieved by purpose-designed systems but sufficient to make the process modification economically attractive under some regulatory regimes.

The main technology for flue gas treatment for nitrogen oxides removal is the selective catalytic reduction (SCR) process developed in Japan and the USA. Its name derives from the fact that it is *selective* in its removal of nitric oxide from the flue gases, leaving other gases unreacted; *catalytic* because it relies on use of catalysts to operate and uses *reduction* to convert the nitric oxide in the flue gases to molecular nitrogen gas, rather than binding it in a chemical compound. Ammonia is interacted with the flue gas nitric oxide in the presence of the catalyst, producing nitrogen and water according to the reaction:

$$4NH_3 + 4NO + O_2 \rightarrow 4N_2 + 6H_2O$$

The catalyst used varies in different systems but vanadium pentoxide is commonly employed. Such systems can achieve 80–90% removal of nitrogen oxides, in conjunction with combustion modifications to reduce nitrogen oxides formation. The process operates only at temperatures in the region of 350–400°C and therefore the unit needs to be installed in the region downstream of the boiler where the flue gases have this temperature.

The economics of flue gas denitrification by the SCR process is critically dependent on the lifetime of the catalyst, which comprises about 25% of the capital cost of such denitrification plants. Early experience showed that catalysts might last for only one year before needing replacement but this area has recently been the focus of intense development work so that expectations of up to three years' lifetime are being made for new systems being installed on plants in Austria and Germany.

Another potential problem with such systems stems from the use of ammonia. If there is any slippage of unreacted ammonia from the denitrification stage, it may cause problems further down the flue gas path besides being obviously an economic penalty. Reaction with sulphur trioxide in the flue gas may cause deposition of ammonium sulphate in sensitive parts of the system, such as rotary air heaters, while if, as is likely, a flue gas desulphurisation system is downstream of the denitrification stage, and it produces gypsum for utilisation, it is undesirable that this be contaminated to any degree by ammonia. In existing plant, however, careful attention to operation seems to have ensured that this problem can be overcome and losses of under 5 ppm v/v of ammonia are routinely quoted.

Experience of SCR operation has been predomi-nantly gained in Japan, although full-scale plants are now under construction in Europe and this recent activity enables the description in the Watt Committee Report No. 14 to be updated. For example, compared with FGD plants, SCR plants do not require particularly large constructions and they have been relatively easily incorporated into plant in the 400 MW(e) range in Europe. Again, compared with FGD plants, they are not expensive in capital or operating costs, with capital costs around $40–50 kW^{-1} and operating costs of 3–6 mills kWh^{-1} (1983).[12] (1 mill = $0.001). However, the routine application of this technology to power stations in Europe still remains to be demonstrated. In the UK, a special problem might be the adverse effects of the alkali and chloride content of indigenous coals on the reduction catalyst.

5.4.4.4 The joint removal of sulphur dioxide and nitrogen oxides

As noted above, some technologies have been specifically designed for joint removal of sulphur dioxide and nitrogen oxides, while others are modifications of systems originally designed for sulphur dioxide removal alone. Those with integrated joint removal capacity essentially exploit the ability of certain sorption agents for sulphur dioxide removal also to act in a catalytic fashion to reduce nitric oxide in the presence of ammonia. The Bergbaufor-schung-Uhde activated carbon system mentioned in Section 5.4.3 above on regenerable flue gas desulph-urisation plants is the only example which is pro-gressing towards early commercial deployment but there has also been experimental work in the USA on a system which uses copper oxide (CuO) as the reagent. Both these systems are complex and use expensive sorption agents. Their economic viability will be very dependent on the attrition and poison-ing rate of the sorption agent/catalyst used and their suitability for wider scale deployment remains to be established.

Several manufacturers of FGD systems have in-vestigated modifications of their original designs to give appreciable levels of nitrogen oxides removal as well. In the USA, the Niro lime spray-drying system has been augmented with 10% w/w of sodium hy-droxide in the input slurry and this, together with some operational modifications, has led to reported nitrogen oxides removal levels of above 50%. The system produces an end product which is a mixture of calcium sulphite, sulphate and nitrate. This, in common with the great majority of spray-drying systems would probably require careful disposal to land-fill.

The Walther FGD system uses ammonia to react with sulphur dioxide to produce ammonium sulphite ($[NH_4]_2SO_3$), which is then oxidised to ammonium sulphate ($[NH_4]_2SO_4$). By injection of ozone into the flue gases, the nitric oxide present can be oxidised to nitrogen dioxide, which also reacts with the ammonia to produce ammonium nitrate (NH_4NO_3). The end product is a mixture of ammonium sulphate and nitrate which can be used in some circumstances as a fertiliser. One such plant has been constructed in West Germany but its reliable, economic operation remains to be demonstrated. Work is also progressing in the USA on the use of electron beams to oxidise the nitric oxide in flue gases to nitrogen dioxide, thereby enhancing its reactivity. One system uses a lime spray dryer to subsequently capture the sulphur dioxide and nitrogen oxides reaction product, similar to the Niro system described above. The other uses ammonia somewhat analogously to the Walther system. Both these processes are at the early stages of development.

5.4.4.5 The use of ammonia in flue gas treatment

Several flue gas desulphurisation systems, and also all selective catalytic reduction systems for nitrogen oxides removal, involve the use of ammonia, as described above. This substance in itself represents an environmental hazard if there were to be any leakage incidents and already in Europe there have been some developments related to this aspect. Local regulatory agencies, in particular, have indicated concern at technologies involving large-scale storage or transport of ammonia and in at least one case in Austria a utility planning to operate a nitrogen oxides removal selective catalytic reduction plant has been required to forego road supply of ammonia for the plant. It is not suggested that the hazards of ammonia utilisation are an insuperable obstacle to deployment of systems using the reagent but the problem indicates that with any form of emission control technology there are always some adverse secondary environmental considerations as well as the additional economic costs of constructing and operating the technologies.

REFERENCES

1. The Watt Committee on Energy, *Report No. 14 — Acid Rain*, Section 4. London, 1984.

2. International Union of Producers and Distributors of Electrical Energy (UNIPEDE), *Experience on desulphurisation of flue gases*, Athens Congress, June 1985.
3. PARKER, L. B. & TRUMBULE, R. E., *Mitigating Acid Rain with Technology; Avoiding the Scrubbing/Switching Dilemma*. Congressional Research Service, US Congress, June 1983.
4. Department of the Environment, *Acid Rain; The Government's Reply to the Fourth Report from the Environment Committee, Session 1983-4*, HC 446-1, HMSO, London, CMND 9397, 1984.
5. Miljoestyrelsen, *Forsuringsudvalget: miljoe og energi*, Miljoeministeriet, Koebenhavn, 1984.
6. HIGHLEY, J., Atmospheric FBC and CFB; the European approach, paper presented at the Sydkraft Energy Research Foundation Symposium on Advanced Coal Power Technology, November 1985. Malmoe, Sweden, 1985.
7. House of Lords Select Committee on the European Communities, *Air Pollution, Session 1983-4, 22nd Report*, HMSO, London, June 1984.
8. House of Commons Select Committee on the Environment, *Acid Rain, Session 1983-4*, Vols. I & II. HMSO, London, September 1984.
9. House of Commons Select Committee on the Environment, *Follow-up to the Environment Committee Report on Acid Rain, Session 1985-6*. HMSO, London, November 1985.
10. BARRATT, G., The economics of clean air, paper presented at the Combined Heat and Power Association annual conference, Torquay, June 1985.
11. DACEY, P. W., *Developments in NO_x control for coal fired boilers*, Working Paper, 67. IEA Coal Research, London, 1984.
12. MORRISON, G. F., *Nitrogen oxides from coal combustion—abatement and control*, Report ICTIS/TR11, IEA Coal Research, London, 1980.

5.5 THE TECHNOLOGY OF CONTROLLED VEHICLE EXHAUST EMISSIONS

5.5.1 Emissions characteristics of internal combustion engines

Passenger motor vehicles in the UK are largely propelled by spark ignition engines with a comparatively few diesel engines: in Germany and other European countries this percentage is higher. Commercial vehicles are almost universally fitted with diesel engines in Europe and are increasing in number in the USA. The pollutant gases from spark ignition engines (SI) are carbon monoxide (CO), unburnt hydrocarbons (HC) and nitric oxide (NO). These pollutants have been considerably reduced to meet successively more stringent EEC regulations and can be reduced still further, albeit with considerable expense. While carbon monoxide and hydrocarbons can be reduced without deterioration of fuel consumption, it is very difficult to reduce oxides of nitrogen without impairing the engine efficiency. HC emissions from petrol vehicles arise also from evaporation and fuelling.

The relative contribution by Road Transport (RT) sources in the UK are shown in Figs. 5.5.1–5.5.4 (taken from the Warren Spring Laboratory Report LR 634 (AP) M). It will be noted that RT produces very little sulphur dioxide as the fuel is well refined; what there is (1%) comes mainly from diesel engined vehicles. However, the major proportion of environmental carbon monoxide (84%) and some 40% of the nitric oxide is from RT. Hydrocarbons, the common term used for this vehicle emission, are about 26% of the total environmental 'Volatile Organic Compounds' (VOC) emission. This latter term is a more accurate reflection of the emission situation as it embraces chemical solvents, alcohols, ketones, aldehydes etc. Figure 5.5.5, from

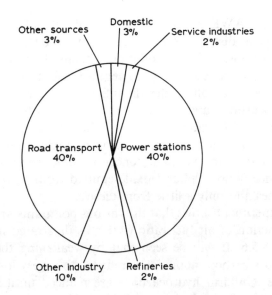

Fig. 5.5.3. UK nitrogen oxides emissions 1985. Total = 1·84 million tonnes NO_2.

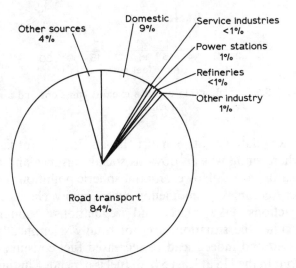

Fig. 5.5.1. UK carbon monoxide emissions 1985. Total = 5·39 million tonnes.

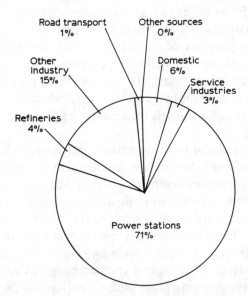

Fig. 5.5.4. UK sulphur emissions 1985. Total = 3·58 million tonnes SO_2.

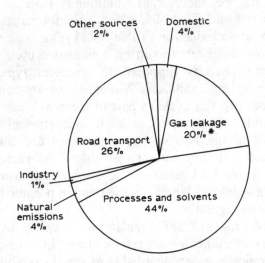

Fig. 5.5.2. UK volatile organic compound emissions 1985. Total = 2·06 million tonnes. The figure for gas leakage (indicated by *) may be slightly over-estimated. The 'gas' is mainly methane, which is only slightly reactive in the environment.

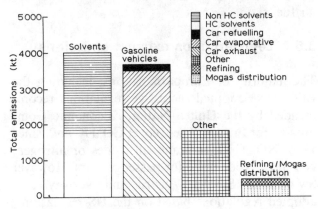

Fig. 5.5.5. Emissions of volatile organic compounds in Western Europe (based on Ref. 1).

a CONCAWE Report[1] for the whole of Western Europe as distinct from the UK, shows the relative proportions. These pollutants themselves are unlikely to cause damage to vegetation but due to chemical reactions under the influence of sunlight, hydrocarbons and nitric oxide together with the oxygen and water vapour in the air undergo complex reactions[2] forming nitrogen dioxide, nitric acid and ozone in small concentrations. There is practically no sulphur dioxide emitted from petrol engines and only a little from diesels.

Experiment shows that the various pollutants are a function of air/fuel ratio (A/F), as illustrated in Fig. 5.5.6. It will be seen that by weakening the mixture carbon monoxide is reduced to a very low value. Unburnt hydrocarbons are reduced until a minimum is reached beyond which HC emissions increase. HC emissions are due to several factors including unburnt fuel being trapped in crevices (e.g. between piston and cylinder/sparking plug) and, particularly as the mixture becomes weak, the slower progress of combustion from the sparking plug with quenching taking place near the combustion chamber walls. If the mixture is further weakened, misfiring would occur, which is quite unacceptable and sets the limit for any given engine design.

There is relatively low nitric oxide formation at a rich mixture, but it increases to a maximum just weak of stoichiometric (the theoretical mixture for complete combustion). Beyond this, nitric oxide reduces as the mixture gets weaker. The reason for this is that nitric oxide is formed from the nitrogen and oxygen of the air reacting in and behind the flame front as it crosses the combustion chamber from the sparking plug. With rich mixtures, there is little oxygen available to form nitric oxide. As NO formation is roughly proportional to temperature, there is a steep fall as mixtures are weakened to 20/1 air/fuel ratio.

5.5.2 Legislation on exhaust emissions

The majority of European nations, including the EEC, have adopted successive regulations recommended by the United Nation's Economic Commission for Europe. These cover CO, HC and, since 1976, NO_x (NO_x indicates all oxides of nitrogen, NO nitric oxide only) for passenger cars. However Switzerland, Sweden, Austria and Norway have adopted regulations based on the US regulations. The EEC regulations have been made progressively more severe in a series of stages since 1970.

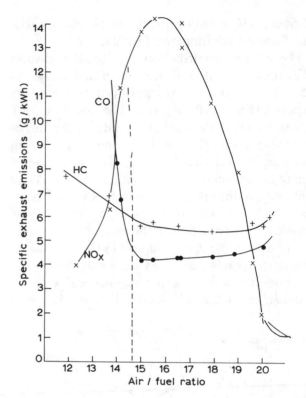

Fig. 5.5.6. Relationship between exhaust emissions and air/fuel ratio.

Regulations of exhaust began in California, where smog was an obvious social nuisance and a health hazard. However, atmospheric pollution is a very complex phenomenon where many chemical reactions take place and regulations, though helping the situation, are not totally scientifically based and indeed lead to increased fuel consumption. In the US at least 5% of fuel is expended in this way.

In Europe, the current Community regulations are based on ECE 15-04, which is the fourth stage of pollution reduction (Table 5.5.1). This requires vehicles under test to run on a simulated cycle of 4·0 km, which is supposed to represent typical urban driving conditions. Modifications are being proposed to this cycle as part of the next stage of pollution reduction to make it more typical of actual driving habits. The regulations are complicated and vary with the weight of the vehicle. Thus Table 5.5.1 relates to light vehicles of below 750 kg weight and indicates the trend in progressive stages of regulation.

New proposed EEC regulations to replace ECE 15-04 (sometimes loosely referred to as 'ECE 15-05' but topically more identifiable as the 'Luxemburg Accord') are less complex and are based upon three categories of engine size on a principle that aims to avoid catalysts for small cars and thus save cost and

Table 5.5.1 EEC exhaust emission limits

Vehicle reference weight (kg)	ECE 15 (1971)		ECE 15.01 (1975)		ECE 15.02ᵃ (1976)	ECE 15–03 (1979)			15–04 (1984)	
	CO	HC	CO	HC	NOₓ	CO	HC	NOₓ	CO	HC + NOₓ combined
Below 750	100	8·0	80	6·8	10·0	65	6·0	8·5	58	19·0

Figures are in grams of pollutant per test.
ᵃ As 15·01 for CO and HC.

Table 5.5.2 Latest EEC exhaust emission proposals (15–05)

Engine Capacity (Petrol engines)	Implementation date new models/new cars	Emission standard (grams/test)
Over 2 litres	1 October 1988/ 1 October 1989	CO = 25 HC + NOₓ = 6·5 NOₓ = 3·5
1·4–2 litres	1 October 1991/ 1 October 1993	CO = 30 HC + NOₓ = 8
Under 1·4 litres (Stage 1)	1 October 1990/ 1 October 1991	CO = 45 HC + NOₓ = 15 NOₓ = 6

fuel wastage but at the same time reduces pollutants by engine design (e.g. 'lean burn') modification. These proposed regulations are shown in Table 5.5.2.

These proposals were agreed under the procedures of the European Single Act at the meeting of Council Ministers on 21 July 1987, with only Denmark dissenting. The proposals therefore pass to the European Parliament, which has three months to examine the agreement. Were the EP to disagree, the Council could only adopt the standards by unanimity.

It should be noted that the Table 5.5.2 emission limits for the 'under 1·4 litres' category of vehicle is deemed 'stage 1', with a second stage proposal for further reduction for this vehicle range which was originally intended to be decided on later in 1987, with implementation proposed not later than 1992 for new models and 1993 for new cars.

EEC emission reductions for diesel vehicles are in train, divided as between (a) passenger cars and related light van categories, with an expected comparability to those of Table 5.5.2 with an added particulate emission clause; however, diesel cars with a capacity between 1·4 and 2·0 litres and direct fuel injection engines need not conform until 1 October 1994 for new models and 1 October 1996 for all production, and (b) commercial vehicles and buses above 3·5 tonnes weight, proposals based on a reduction at a number of levels from the existing

ECE R49 values are still in discussion within the Commission, taking into account future diesel fuel quality and future engine design, and with, again, stricter control on particulate emission.

5.5.3 Approaches to exhaust emission reduction

The reduction of pollutants is possible by the adoption of either or both of two approaches. The first is to use catalysts to convert combustion products to harmless substances. Oxidation catalysts can be used to oxidise hydrocarbons and carbon monoxide. The more complex three-way catalysts are capable of reducing the three major pollutants, but require a very sophisticated fuel control system. At present they are the only way of meeting the US regulations, and the new European Regulations (15-05) will require this method for vehicles with engines over 2 litres in capacity. Catalysts are discussed in Section 5.5.3.3.

The second approach relates only to nitric oxide and the principle is to reduce the temperature of combustion. In general, with conventional engines, this is accompanied by loss of efficiency. Two methods are used. The first, ignition retardation, reduces the maximum temperature of combustion and consequently the NO. This is illustrated in Fig. 5.5.7 for a series of mixture strengths. Because of the reduction in efficiency it is only used if essential

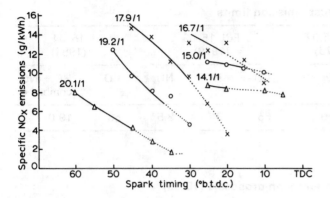

Fig. 5.5.7. Effect of mixture strength and spark timing (degrees before top dead centre) on specific exhaust emissions, for a series of air/fuel ratios.

to meet regulations. The preferred method is exhaust-gas-recirculation (EGR). With this system a small amount of exhaust gas (say up to 10%) is introduced into the inlet manifold and dilutes the charge. This reduces the temperature of combustion (due to its relatively high specific heat) as with lean-burn, but also avoids increasing the oxygen content of the combustion gases.

However, it is possible to design spark-ignition engines in which the efficiency may be preserved with lower nitric oxide emissions. These engines endeavour to utilise the advantages of the diesel engine without the disadvantages of the expensive fuel injection equipment and heavier weight, which in its turn has a fuel consumption penalty. The method is to produce engines that take in a lot more air. These engines can be divided into two classes: lean-burn engines and stratified-charge engines.

5.5.3.1 The lean-burn engine
We will consider the lean-burn approach first. Such engines endeavour to reduce emissions through burning weak mixtures which are homogeneous in character, that is, the A/F ratio is as constant as one can achieve. If a large reduction in nitrogen oxides is required, the A/F ratio needs to be about 20/1. Unfortunately, as the mixture is weakened the rate of burning (heat release) becomes slower and generally speaking there is a loss of efficiency. However, by raising the compression ratio and increasing turbulence in the combustion chamber the speed of burning can be increased again, but unfortunately high compression ratios cause a greater tendency towards 'knock'. Knock is familiar to people who can remember the low-octane fuel days during and following the Second World War when most engines 'knocked' at low speeds and high loads, namely when accelerating at full throttle, or

climbing hills. This type of knock tends to die out with high speed. However, high-speed knock occurs in some, but not all, engines. It is possible to prevent this knock by ensuring that the ignition setting is prevented from advancing into the sensitive areas, but clearly this requires much more precise ignition equipment than is commonly used. However, such equipment is indeed available and is being progressively introduced. Because of the great reduction and, ultimately, elimination of lead from fuel the Octane No. (Research Method) of premium fuel is likely to drop from the present 98 Octane to 95 and this means a compromise in the compression ratio (CR) selected. A modest CR of 9/1 could be used with a 95 Octane fuel with only a small retardation at low speeds on full load. At 12/1, about the optimum ratio for fuel economy, considerable retardation is necessary at low speeds and some retardation at high speeds, for near full-load operation. However, this may still be the preferred method for small- and medium-sized engines to meet the new European Community emission standards. The reason for preferring this is that at part load, under cruising conditions, knocking is not a problem and some loss of power and economy near full load is tolerated.

The design details of lean-burn engines have to balance the concerns of nitrogen oxide and hydrocarbon emission and fuel economy. The situation is shown in Figs. 5.5.8 and 5.5.9, where pollutant concentrations are plotted against A/F ratio. The results shown are for an experimental engine running under cruise conditions with two types of cylinder head, one with a four-valve hemispherical combustion chamber (CC) and the other with a

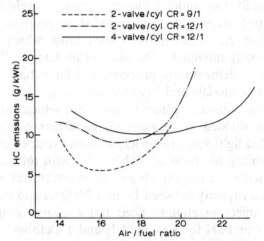

Fig. 5.5.8. Relationship between hydrocarbon emissions and air/fuel ratio for an experimental 500 cm³, single-cylinder petrol engine (2000 rpm, 2 bar BMEP).

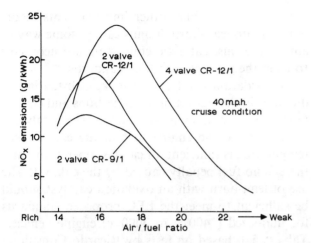

Fig. 5.5.9. Relationship between nitrogen oxide emissions and air/fuel ratio for an experimental engine.

two-valve biscuit-shaped CC. It will be noted, however, that with a four-valve cylinder head the better filling gives an even worse NO_x situation near stoichiometric A/F but combustion can be continued to a weaker A/F ratio where low NO_x is obtained.

Due to the high induced turbulence with the four-valve cylinder head fast burning is achieved with improved fuel consumption. The hydrocarbon emissions are numerically higher with the four-valve but can be maintained to a considerably weaker A/F ratio. Figure 5.5.10 shows the balance of economy against pollution concentration. HC are added to NO_x as in the current EEC regulations. A/F ratios are shown against the concentration of pollutants. It will be seen that the lowest pollutants are obtained for the two-valve 9/1 CR engine but this is at a considerable fuel penalty. A much improved fuel economy, with little sacrifice in pollution concentration, is obtained with the four-

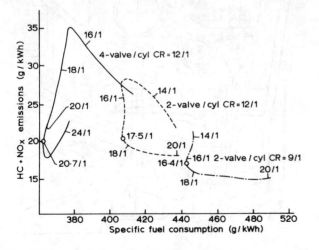

Fig. 5.5.10. Emissions and fuel consumption for an experimental engine.

valve engine 12/1 CR, providing it can operate satisfactorily at about 20-22/1 A/F ratio.

The pollution reductions described above are for a single-cylinder research engine. For the practical multi-cylinder situation major problems are encountered because engine power is reduced at high A/F so that a mixture nearer to stoichiometric is necessary for achieving adequate vehicle acceleration and maximum speed. Despite some difficulties lean-burn engines are of major interest to European manufacturers and are expected to be able to satisfy the prospective regulations EEC 15-05 for the smallest engine sizes (below 1·4 litres).

5.5.3.2 Stratified charge engines
In an endeavour to overcome the limitations of homogeneous lean mixtures, several types of stratified charge engines have been investigated. Stratified charge engines are designed to have a near-stoichiometric A/F ratio near the sparking plug but weakening off in the rest of the combustion chamber. These engines may broadly be divided into two classes, the open chamber and the pre-chamber types. This is a similar division to the diesel engines, where we have open chamber DI (direct injection) and pre-chamber IDI (indirect injection) systems. Both systems have their advantages and disadvantages. The open chamber stratified charge engine has been well demonstrated in the Ford Proco and Texaco system (illustrated in Fig. 5.5.11). The Texaco system operates somewhat like the open chamber diesel engine but with a spark to initiate the combustion while the Ford Proco contrives that a near-stoichiometric mixture is present in the vicinity of the sparking plug for ease of ignition and that the rest of the chamber has a lean A/F ratio. Neither of the systems has reached large-scale production, due mainly to their extra cost of manufacture.

It will be appreciated that there is great difficulty in stratifying the charge in an open combustion chamber in a desirable and controlled way under all conditions of loads and speeds, and this difficulty can to some extent be ameliorated by having a separate pre-chamber with two separate fuel supplies. Again there are two broad types of pre-chamber engines. There are those which have a third valve, exemplified in the Hondo CCVC engine,[3] which was based on principles and early experimental work by Sir Harry Ricardo. These engines have two carburettors, one supplying the main chamber, and the second small carburettor

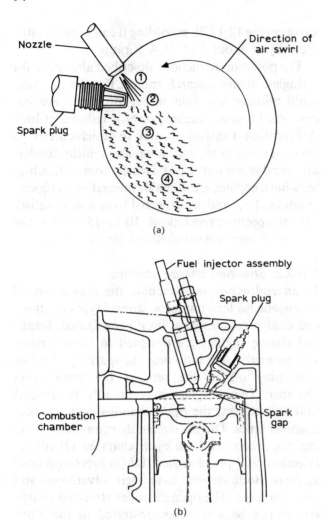

Fig. 5.5.11. Open-chamber stratified charge designs. (a) Texaco combustion process: 1, fuel spray; 2, fuel–air mixing zone; 3, flame front area; 4, combustion products. (b) Ford 351-CID PROCO engine—cross-section.

supplying the pre-chamber through the third valve. The second type has the usual two valves and fuel is injected into the pre-chamber by an injection pump system. Two carburettors are generally less expensive than fuel injection equipment. British Leyland[4] has done considerable experimental work with the three-valve engine (see Fig. 5.5.12). The mode of combustion is very different in these engines from that of the normal engine in that a relatively rich charge (stoichiometric) is burned in the pre-chamber and is thus easily ignitable and this sends a jet of flame into the main chamber which ignites the lean mixture. However, heat losses are greater with pre-chamber engines giving slightly lower efficiency.

5.5.3.3 Catalytic systems
The lean-burn approach favoured by many European manufacturers reduces nitrogen oxide emissions but can lead to an increase in hydrocar-bon emissions. Whilst other improvements incorporated into lean-burn designs can go some way to ameliorate this, catalytic convertors are necessary to meet the more demanding standards.

An oxidation catalyst can make substantial reductions in the output of hydrocarbons and carbon monoxide from a lean-burn engine. The catalyst is fitted close to the engine so that the exhaust gas temperature is sufficient for catalyst operations, and there is no fuel penalty caused by the catalyst. The use of lean-burn with an oxidation catalyst should be sufficient to meet the EEC proposed standards for mid-sized (1400 cc to 2000 cc) engine vehicles. Table 5.5.3, based on tests by Ricardo Consulting Engineers on a Toyota Carina vehicle,[5] illustrates this.

To achieve the current US standards, or the proposed EEC standards for cars over 2 litres, three-way catalytic systems are used, so called because they reduce hydrocarbons, carbon monoxide and nitrogen oxides simultaneously. The first two are converted by simple oxidation to carbon dioxide and water. Nitric oxide is reduced by the carbon monoxide and hydrogen present when an engine is running with a stoichiometric mixture, or richer. The situation is shown in Fig. 5.5.13. The reactions at stoichiometric air/fuel ratio are as follows:

$$2C_8H_{18} + 25O_2 \rightarrow 16CO_2 + 18H_2O$$

$$2CO + O_2 \rightarrow 2CO_2$$

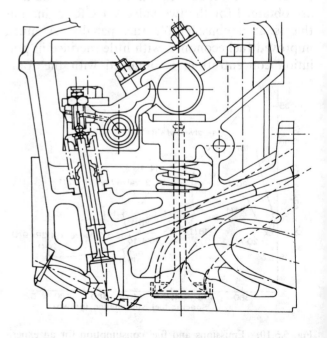

Fig. 5.5.12.   British Leyland stratified charge research engine.

Table 5.5.3 Comparison of achievable emissions with proposed standard ECE 15–05 for a mid-engine size car (Toyota) using an oxidation catalyst

	HC	NO_x	$(HC + NO_x)$	CO
Proposed EEC standard (1·4–2 litres)			8.0	30.0
Without catalyst	9·7	5·52	14·69	20·73
With oxidation catalyst	1·80	4·45	6·25	9·32

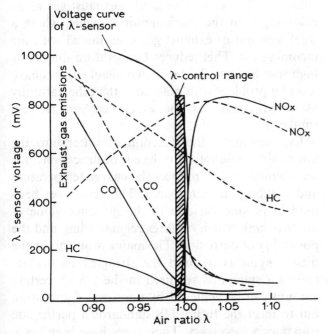

Fig. 5.5.13. Characteristic of zirconia λ-sensor and exhaust gas concentrations with (——) and without (– – –) catalyst.

$$2NO + 2H_2 \rightarrow 2H_2O + N_2$$

$$2NO + 2CO \rightarrow 2CO_2 + N_2$$

Very fine mixture control is required to ensure that oxygen, carbon monoxide and hydrogen are all present. This requires feed-back control utilising a 'sensor', monitoring oxygen potential in the exhaust gases and adjusting the fuel system as necessary. Some form of electronic control is necessary for precise metering of the fuel. This can be electronic-ally controlled fuel injection to individual cylinders, or an electronically controlled carburettor, or so-called throttle body injection. Electronically con-trolled ignition is also desirable.

The need for this control equipment adds a con-siderable cost penalty, and the stoichiometric running represents a loss of fuel economy of about 5% since maximum engine efficiency is obtained with leaner mixtures.

Results of tests performed by Johnson and Matthey[6] using a VW Scirocco (1.8 litre engine capacity) are shown in Table 5.5.4. The catalyst was previously run for 50 000 miles at high speed. The efficacy will be apparent.

5.5.4 Hydrocarbon evaporative emissions

The emission of hydrocarbons considered above are those emitted from the exhaust. However, a not-inconsiderable amount of VOC from road vehicles comes from other sources: first, from the fuel tank by evaporation, secondly from other parts of the fuel system, e.g. float chambers, and finally from upholstery paints and glues used in manufacture. Figure 5.5.5 above shows the contribution by road transport for Western Europe.

There are ways of reducing these emissions. It is already mandatory in Europe to seal the fuel system and engine crankcase and to vent these to the intake of the engine, from whence they are subsequently burned.

Further reductions are possible, and the relevant

Table 5.5.4 Comparison of emissions with proposed standard ECE 15–05, for a mid-engine size car (VW) using a three-way catalyst

	HC	NO_x	$HC + NO_x$	CO
Proposed EEC standard (1·4–2 litres)			8·0	30·0
Without catalyst	4·5	5·0	9·5	43·0
With three-way catalyst	1·4	1·05	2·45	16·6

All figures are in grams of pollutant per test.

technology is the onboard carbon canister, endorsed by the US EPA and fitted over a period of 10 years or more as a fixture to the fuel tank system of the automobile. It is able to absorb some 90% of the evaporative losses that occur as normal running losses, 'hot soak' losses when the car is idle, and 'diurnal' loss due to daily temperature variation and consequent breathing. The vapour held on the carbon is desorbed during normal running by the engine air intake and so ends up as forming a part of the fuel input. There are two forms of canister; the first is sized to cater for the evaporative onboard running losses referred to above, while the second is an enlarged canister able also to absorb the much smaller 'refuelling' loss that occurs while the vehicle is being fuelled. Either makes a significant contribution. The US style 'large' carbon canister would cost between $20–80 per vehicle.

The projections shown in Fig. 5.5.14, based on the CONCAWE report referred to above,[1] are for hydrocarbon emission from petrol driven vehicles. They show the effect that EEC controls are likely to have in reducing emissions, and also the further reduction that could be available should large canisters be required by regulation.

5.5.5 Diesel engines

Diesel engines are used to power most commercial vehicles and buses and, in Europe, a proportion of private cars. These also have pollution problems, the most difficult being particulate emissions. There are European regulations for carbon monoxide and for hydrocarbons and nitrogen oxides combined. The diesel, because of the excess air, has no problem with carbon monoxide. Small engines (below 3 litres) can meet the combined nitrogen oxide/hydrocarbon regulations without much difficulty but large commercial vehicles have some difficulty here.

Two palliatives are available but both have disadvantages. The first is to retard the injection timing, which reduces the maximum temperature of combustion and hence nitrogen oxides but also reduces efficiency. The second method is exhaust gas recirculation, as in the spark ignition engine, where a small amount of exhaust gas is entrained with the incoming air. This reduces temperature due to its high specific heat. Turbo-blown diesel engines have less of a problem with pollutants than the naturally aspirated engine as the excess oxygen reduces smoke.

Experimentally the injection of water into the intake air, preferably into the eye of the compressor on a blown engine, reduces the charge temperature and reduces nitrogen oxides. This has never been used in production due to the nuisance value of another tank which requires regular filling and the possibility of corrosion. The major problem for the diesel engine is to meet the stringent particulate emission standards required in the US. A certain amount can be achieved by improved combustion but to meet the future US standards a particulate trap may be necessary. These traps have been made and tested experimentally but have not yet proved good enough for production. There are two types, those that work through filtration of the exhaust and those that catalytically oxidise the particles.

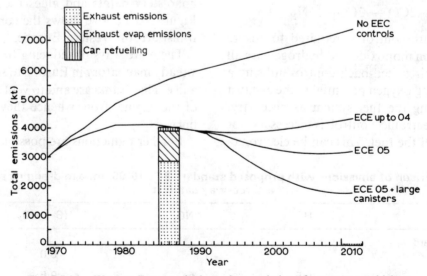

Fig. 5.5.14. Western Europe—hydrocarbon emissions from motor vehicles.

5.5.6 Costs

Costs are difficult to estimate and must take account of many factors. They will depend upon whether a system is being incorporated into a new or an existing model. For example, to house a catalyst in an existing vehicle, a space has to be made which may alter pressings and require retooling. At least in the shorter term, costs of emission control may be only poorly reflected in showroom prices, which will depend upon a manufacturer's marketing strategy. Overall costs need also to include changes in fuel economy and maintenance requirements. Some approximate costs have been obtained from a manufacturer. These are ex-works costs without VAT, special car tax or retail trade profit. No attempt is made to establish the current extra cost to meet the 15-04 regulations as the additional features to reduce pollutants have been designed to improve fuel consumption as well. Estimates of capital cost to meet the proposed 15-05 regulations over and above that required at present are for the lean-burn combustions(for small engines) £50; lean-burn plus oxidising catalyst (for mid-range vehicles) £200; for three-way catalysts plus feed-back, £380.

5.5.7 Summary

There are many methods of reducing the acidic gases emitted from internal combustion engines but the expense of equipment increases considerably as the extent of reduction is increased. The proposed EEC regulations are a compromise between severity, cost and fuel economy. Further research and development will be required and considerable road running before the final preferred methods are adopted. A probable outcome will be a lean-burn approach for vehicles in the small range (engines below 1400 cc) and, in the medium range, a lean-burn approach with an oxidising catalyst (or perhaps a three-way catalyst system), while for larger vehicles the three-way catalyst with feed-back will almost certainly be utilised.

Acknowledgements

The experimental data in this paper have been taken with permission from publications of Shell and British Leyland, which are gratefully acknowledged. The results quoted are typical of similar work performed at other motor and oil companies.

REFERENCES

1. CONCAWE, Report No. 87/60.
2. REYNOLDS, S. D., ROTH, P. M. & SEINFIELD, Mathematical modelling of photochemical air pollution. *Atmos. Environ.*, **7** (1973) 1033–61.
3. YAGI, S., DATE, T. & INOUE, K., NO_x emissions and fuel economy of the Honda CVCC engine. SAE Paper 741158.
4. WEAVING, J. H. & CORKILL, W. J., British Leyland experimental stratified charge engine. Paper C246/76 presented at Institute of Mechanical Engineers Conference, Nov. 1976.
5. EVANS, W. D. J. & WILKINS, A. J. J., paper presented at Second International Symposium on Highway Pollution, 1986.
6. EVANS, W. D. J. & WILKINS, A. J. J., Single bed, three way catalyst in the European Environment. SAE Paper 852096.

5.6 MONITORING IMPLICATIONS OF EMISSION CONTROL PROPOSALS

5.6.1 Background

Monitoring is an inescapable part of any emission control strategy, and is an important part of most air pollution control legislation. The proposed EC legislation to control emissions from large combustion plant, COM(83)704 and its update COM(85)47, is no exception. The draft directive seeks to control the emissions of sulphur oxides (SO_x), nitrogen oxides (NO_x) and particulate material from plants over 50 MW thermal input. It calls for an initial bubble reduction of these emissions from a chosen datum year of 1980. It then stipulates emission limits for specified categories of plant and fuel. Deciding the magnitude of the bubble reduction has been a contentious issue because in 1980 there was little or no monitoring being carried out in many member states: consequently estimates of emissions at that time are only as good as the emission inventories and emission factors used to calculate them. The data for sulphur dioxide emissions is reasonable because sulphur in fuel is measured by fuel suppliers but estimates for NO_x and particle emissions are less good because the emission factors used there are highly plant-dependent and can be very variable.

These inherent uncertainties highlight the differences between the two alternative approaches which can be adopted for controlling the implementation of emission control legislation: either the use of measurement or of more comprehensive emission factors by which the emissions from plant would be deduced from the composition of the fuel used and the type of plant and combustion conditions. The Commission opted for measurement and the draft directive not only calls for all new plant to be fitted with monitoring instrumentation but also requires retrofitting of instruments to existing plant with an expected life of over 10 000 operating hours. The monitoring requirement, based very much on German experience, would produce data with a high statistical significance. Both the accuracy and resolution required are high; while accuracy is obviously desirable the need for high resolution is more arguable. The response of acid deposition phenomena to emissions is relatively slow but NO_x and particulate emissions can be sufficiently variable that high resolution measurements may be needed if significant excursions above the control limits are to be avoided. However, the half–hourly resolution specified, which is a compromise between the performance of modern instrumentation and practical calibration considerations, creates a response time requirement of any abatement strategy which may be difficult to accommodate; controlling the sulphur content of fuels to the levels required would be costly.

To check compliance with the emission limits set in the draft directive it will be necessary not only to monitor NO_x, SO_x, and particulate concentrations but also to monitor the oxygen and water concentrations prevailing at the point of monitoring so that the measurements can be compared on a common basis under standard conditions. Hydro-

carbon emissions are not covered by this draft directive but may be subject to controls in the future.

At the moment there are considerable differences between the emission regulations of the member states which reflect differing attitudes to pollution control: some are much more comprehensive than others. Similarly there are many measurement methods used to ensure compliance, see Tables 5.6.1–5.6.3, and the draft directive aims to harmonise these. The monitoring requirements specified to police the draft directive are:

(a) SO_x, NO_x, and particulates should be measured continuously but some leeway is allowed for individual measurements of SO_2;
(b) half-hourly and daily means must be calculated for continuous measurements;
(c) means should be corrected to a 3% oxygen content of the flue gas for liquid and gaseous fuels and 6% oxygen for solid fuels at 273 K and 1013 mbar;
(d) the data should be stored not only as a frequency distribution which must be readable at any time, but in such a form as to be able to show compliance with the requirements of the directive; that is, none of the daily means should exceed the emission limit values, 97% of the half-hourly means should not exceed 1·2 times the limit value and none of the half-hourly means should exceed twice the limit value;
(e) monitors should have an availability greater than 80%;

Table 5.6.1 Survey of methods and guidelines for particulate measurement applied in member states

Member state	Discontinuous	Continuous		
	Gravimetric	Photometric in-situ	Light scattering	Beta ray absorption
Belgium	NBN × 44–002	×		×
Denmark	×[a]			
Germany	VDI 2066 Parts 1 & 2	VDI 2066 Part 4 (draft)	×	×
Greece	×			
France	AFNOR/NFX 44 052		×	AFNOR/NFX[b] 43017
Ireland	×			
Italy	×	×		
Luxembourg	×			
Netherlands	NPR 2788	×		
UK	BS 893 BS 3405			
Portugal	×	×		
Spain	×	×	×	×

[a] ×, Method is used but there is no national guideline.
[b] Air quality method applicable to source measurement.

Table 5.6.2 Survey on methods and guidelines used for SO₂ measurement in member states

Member state	Discontinuous/manual				Continuous/automatic										
					Extractive								In-situ		
	Iodometric thio-sulphate	Titrimetric hydrogen peroxide	Gravimetric hydrogen peroxide	Hydrogen peroxide-thorin[a]	Non-dispersive infra red	Non-dispersive ultra violet	Electro-chemical	Ultra violet fluor-escence	Conducto-metrical	Flame photo-metrical	Inter-ferential	Non-dispersive infra red	Non-dispersive ultra violet	Dispersive ultra violet	2nd derivative ultra violet
Belgium				NBN T95–202			×			×					
Denmark						×						×			
Germany	VDI 2462 Part 1	VDI 2462 Part 2	VDI 2462 Part 3	VDI 2462 Part 8	VDI 2462 Part 4				VDI 2462 Part 5				×	×	
Greece															
France	×				NFX 20351			NFX 43019[c]	× NFX 20355	NFX 43020[c]	×				
Ireland	×	×		×											
Italy															
Luxembourg															
Netherlands	NEN 3104		NEN 3104												
UK	×	×		BS 1747	×	×		×				×		×	×
Portugal			×	×	×										
Spain				×	×			×							×

Belgium Gravimetric: ×[b]

[a] ISO method.
[b] × Method is used but there is no guideline.
[c] Air quality method applicable to source measurement.

Table 5.6.3 Survey on methods and guidelines used for NO$_x$ measurement in member states

Column groups: *Discontinuous/manual* = Phenol-disulphonic acid, Acidimetric titration, Sodium salicylate, 2,6 dimethyl phenol, Chromotropic acid, Saltzman. *Continuous/automatic — Extractive* = Non-dispersive ultra violet, Dispersive ultra violet, Non-dispersive infra red, Chemi-luminescence, Inter-ferential, Non-dispersive ultra violet. *Continuous/automatic — In-situ* = Dispersive ultra violet, Non-dispersive infra red, 2nd derivative ultra violet.

Member state	Phenol-disulphonic acid	Acidimetric titration	Sodium salicylate	2,6 dimethyl phenol	Chromotropic acid	Saltzman	Non-dispersive ultra violet (Extr.)	Dispersive ultra violet (Extr.)	Non-dispersive infra red (Extr.)	Chemi-luminescence	Inter-ferential	Non-dispersive ultra violet (Extr.)	Dispersive ultra violet (In-situ)	Non-dispersive infra red (In-situ)	2nd derivative ultra violet
Belgium			×[a]		NBN T95-301		×			×					
Denmark															
Germany	VDI 2456 Part 1	VDI 2456 Part 2	VDI 2456 Part 8 (Draft)	×			×	×	VDI 2456 Part 3	VDI 2456 Part 7			×		
Greece															
France										NFX 43018[b]					
Ireland															
Italy										×					×
Luxembourg									×	×					
Netherlands			NEN 2044 (Draft)							×					
UK			BS 1747			BS 1756 Part 4			×	BS 1747				×	
Portugal									×						×
Spain	×														

[a] × Method is used but there is no guideline.
[b] Air quality method applicable for source measurement.

(f) the reliability and calibration of continuously operating measuring equipment must be checked at regular intervals.

By testing legal compliance using instrumentation it will be essential that internationally agreed standardised instrument suitability tests are established. These tests would need to be carried out both in the laboratory and the field by accredited organisations, tests carried out in one member state would have to be recognised in the others. Calibration and maintenance work would require specially trained staff. Since it will be costly and time-consuming to establish agreed criteria there is an urgent need to assess requirements now if UK industry is not to be put at a disadvantage should the draft directive or something similar be adopted.

5.6.2 Monitoring

In a recent international study[1] on the monitoring implications of the draft directive it was concluded that for each pollutant one manual chemical method, preferably an ISO standard, should become the primary reference for continuous measuring devices. Of the several SO_2 methods in use in Europe, the hydrogen peroxide-thorin method is widely accepted, is a reference method in the USA and will shortly be published as an ISO reference method. The preferred method for NO_x analysis would use sodium salicylate, which is currently used in a number of states, but the time lapse of 24 hours between sampling and obtaining the results is undesirable. Two other methods, one based on chromotropic acid and another on 2,6,dimethyl phenol, have recently been developed for more rapid analyses but their precision compared with sodium salicylate has not yet been fully evaluated. An ISO standard for particulate measurement has long been under consideration but considerable work still remains to be done.

5.6.2.1 Particulate monitoring

In-stack transmissometers and Beta ray attenuation instruments are the two most widely used particulate analysers. Transmissometers have the advantage that there is no physical disturbance of the gas stream and the measurement is made over a relatively large gas volume. But the instruments are sensitive to maladjustment of the light beam and careful positioning is required to overcome these problems. The main disadvantage, however, is that the calibration is based on the assumption of constant particle characteristics. Beta ray attenuation instruments give results which are virtually independent of particulate composition and size but they require a sophisticated sampling system to collect representative samples and transfer them intact to the sensor. The difficulties in the design and operation of such systems should not be underestimated. All methods for particulates have the disadvantage that parameters such as attenuation or light scattering are used rather than actual particle mass. Therefore greater uncertainties have to be expected than are usual in the measurement of gaseous components. The range of uncertainty given by the confidence and the tolerance intervals differ from plant to plant but may be as high as 30–50% of the given standard for optical methods.

5.6.2.2 Sulphur dioxide

Of all the available continuous methods, photometric gas analysers are by far the most common and the ones best able to meet the requirements of the draft directive. At present non-dispersive infra red (NDIR) instruments outnumber the non-dispersive ultra violet (NDUV) but the latter may be less prone to water vapour interference. Extractive instruments are more common than in-situ devices and more reliable performance data are available for them. In general extractive instruments appear to have greater precision and stability than existing in-situ instruments, some of which have problems with cross-sensitivity and sensitivity to misalignment. However, cross stack instruments have the inherent advantage that they linearly average over their measurement path, although in cases of severe stratification obtaining a representative measurement can still be a problem. They also have the advantage that one analyser can be used for several analyses. There can be problems with sample conditioning when using extractive devices, particularly in removing water vapour if the analyte gas is soluble.

Extractive on-line monitors are often more cost-effective for existing plant but, for new plant specifically designed to incorporate emission measuring equipment, in-situ devices are likely to be favoured in the future. In general in-situ gas analysers are at an earlier stage of development than extractive instruments, but they have considerable advantages providing that an acceptable quality of measurement can be achieved. This could be a profitable area of new business for the British companies who are in the forefront of in-situ instrument design and development.

5.6.2.3 Oxides of nitrogen

For oxides of nitrogen chemiluminescence methods are widely used, have a good track record and are specified in national standards of several member states. But UV/IR photometric instruments, both extractive and in-situ, are available and have the same advantages and disadvantages of those for SO_2 analysis. In general instruments for nitrogen oxide measure the monoxide only, but this is acceptable providing that the amount of NO_2 does not exceed 5%.

5.6.2.4 Oxygen and carbon dioxide

Oxygen or carbon dioxide are measured at all large plants for combustion control, but since they are usually located upstream of any abatement equipment it would be necessary to install additional instrumentation since measurement should be made in the stack to ensure conformity. Here too, extractive and in-situ instruments are available.

5.6.2.5 Volume flow, temperature and humidity

These are essential components of emission monitoring and the methods used are similar in all member states. Temperature measurement is not a problem but the instrumentation for continuous measurement of volume flow at elevated temperatures is poorly developed/demonstrated except for well-established but limited methods, such as pitot tubes, etc. It is possible to calculate the volume flow from the fuel characteristics for many large combustion processes but not for installations such as cement kilns, etc. Consequently more work is required on the development of this type of instrumentation.

5.6.2.6 Data treatment

The large volume of data resulting from monitoring would have to be treated by specialised collection devices capable not only of storing large volumes of data but also of computing emissions under standard conditions and carrying out limited statistical manipulation on derived data. However, there should be a 'spin off' from using data-loggers of this type because, for little extra cost, they should be able to provide operators with valuable process control information derived from the monitoring data.

5.6.3 The scope of the draft directive

The total number of UK plant over 50 MW likely to be affected by the draft directive is not known with

certainty, although some reasonable estimates have been made by the University of Sussex Science Policy Research Unit (SPRU). Information on industrial boilers is not collected on a systematic basis and beside the problem of identifying the plant there is the added difficulty of differentiating between boilers and installations. The latter point is very important when estimating the costs of instrumentation which it would usually be desirable to fit to individual boilers rather than stacks, even where several boilers share a common stack: in the event of emission problems operators will want to reduce shutdowns to a minimum. The absence of comprehensive statistics makes detailed analysis and costing impossible and some effort should be made to collect further information in this area.

5.6.4 Monitoring costs

Table 5.6.4 shows the current order of costs for the instrumentation. These are to some extent dependent on the measurement principle of the instrument and future prices are likely to reflect market forces as well as manufacturing costs.

The overall cost of fitting the instrumentation, providing access platforms and support services such as power and lighting, etc., could be considerable and in many cases would be higher than the costs of the instrumentation alone. The cost of fitting instrumentation also depends upon the configuration of the plant, but the not infrequent need to make measurements on several boilers on the same plant might allow a common set of extractive

Table 5.6.4 Investment costs

	Measurement principle	Investment cost of one instrument (£ thousands)
Particulates	Opacimeters ⎱ Transmissometers ⎰	1·6–7·0
	Beta attenuation analyser	11·0–15·0
	Multicomponent dust/SO_2	15·0–19·0
Gaseous components	Extractive (NO, SO_2 and O_2)	19·0–27·0
	In-situ Photometers (NO and/or SO_2)	10·0–12·0
	Zirconia cell/probe (O_2 only)	1–2·5
Data collection		3·2–25

instruments to carry out all the measurements by switching between separate sampling lines.

The example illustrated in Fig. 5.6.1 covers the hypothetical case of retrofitting five instruments (O_2, NO_x, SO_x, H_2O, and a transmissometer). The estimated total cost was of the order of £250 000. This case is extreme and often costs would be less, but it provides a useful upper range figure for costing purposes. On this basis the total cost of equipping all UK plant could be between £100 million and £200 million. The cost for a plant would be proportional to the number of boilers involved but, as mentioned above, the layout of each plant and the construction of the ductwork/chimney involved would make the cost of monitoring highly site-specific. The cost of fitting instrumentation on a new plant designed from conception to cater for monitoring would be much less than retrofitting.

Data on operational expenses is sparse but the costs of running and maintaining the instrumentation with the high degree of availability required could be considerable. In the UK, one manufacturer using a large number of monitors reported[2] that 25% of instruments were faulty when received from the manufacturer, a six-week debugging period was necessary before a new instrument could be used on site, but instruments were then expected to have an eight-year useful life. Also:

— most faults were encountered with the sampling systems rather than the actual instruments;
— one man could look after £150 000 of installed capital cost (FGR claim manpower require-

ments of between 20–60 instruments man^{-1}year^{-1});
— annual maintenance costs of approximately 23% of installed capital cost;
— spur sampling lines cost approximately 20% of installed capital costs;
— for extractive IR instrumentation the installed capital cost of £15 000 with a maintenance time requirement of 108 h year^{-1} and a maintenance cost of £1 600 year^{-1};
— for extractive UV the installed capital cost was again £15 000 with maintenance time of 199 h year^{-1}, and maintenance costs of £3000.

The cost of instrumentation as a percentage of throughput/product is likely to be higher for smaller plant than large ones, especially in some industrial sectors.

5.6.5 Conclusions

Source monitoring has been put forward as the way of policing the draft EC directive on the control of emissions of SO_x, NO_x, and particulates from large combustion plant. Consequently it is important that the monitoring methods used by member states are comparable and ISO standards should be developed as primary standards. A number of instrumental techniques exist which will meet the monitoring requirements but some promising new techniques are expected to be more cost-effective on new plant. Similarly, the draft directive calls for some computation on the logged data. Here again, suitable data loggers do exist but there is considerable scope for the development of new models which will also provide plant operation data. The development of this equipment could provide an opportunity to the UK instrument industry. The total costs of meeting the monitoring requirements could be high; there are two components: capital and maintenance. The capital cost of purchase, installation and access are highly site-dependent but could be as high as £250 000 in the case of a difficult retrofit. The cost of maintenance is likely to be in the order of 25% of the installed capital cost per annum.

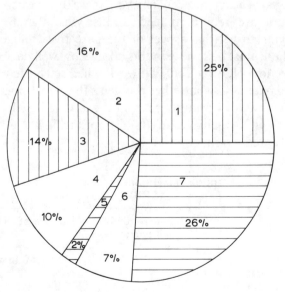

Fig. 5.6.1. Implementation costs: 1, analysers; 2, labour OHD; 3, electrics; 4, structures; 5, services; 6, scaffolding; 7, contingency.

REFERENCES

1. BRIEDA, F., BULL, K., BUHNE, K. W., CALLAIS, M., MENARD, T., WALLIN, S. C. & WOODFIELD, M. J., Methods of sampling and analysis for sulphur dioxide, oxides of nitrogen and particulate matter in the exhaust gases of large combustion plant. Submitted to the Commission of the European Communities, Oct. 1985.

2. DAVIS, A. A., The economics of maintenance. Paper presented to the Institute of Measurement and Control at the discussion meeting of the Industrial Analytical Instrumentation Group, 25 Feb. 1986, London.

5.7 IMPLICATIONS FOR INDUSTRY

Most discussion about emission control, particularly in the British context, has focused upon power generation. This is appropriate, since it is here that the largest reductions will have to be made. However, although the contribution from industry in aggregate will be small, from the point of view of individual industrialists the issue is an important one. Emission control could add to costs in an extremely competitive environment, and prospective legislation represents a source of uncertainty in industry's planning. This paper explores emission control from an industry perspective.

The implications of emission control for industry have to be set against a background of considerable uncertainty. We do not know what the final form of any EEC Directive will be and in particular what combustion plants other than boilers will be included. Within that we do not know what any UK policy for emissions from existing industrial plant may be. No complete register of existing industrial boilers exists nor is there a clear view on availability and price of key fuel elements. We also hear much of the technical developments but nobody knows for certain how they will work out, particularly when applied to the multiplicity of existing boiler stock.

In order to reduce these uncertainties somewhat, let us look at some best estimates. The first thing to say, however, is that the life of industrial boilers in the range covered by the proposed EEC Directive (over 50 MW) is very long, typically 40 years or more. Should legislation relate to existing, rather than only new, boilers, this would have a much more serious implication for industry.

Table 5.7.1 gives information about the existing boiler stock—the estimated numbers of boilers above and below 50 MW and their respective emissions of sulphur dioxide and nitrogen oxides. The data are from early work by SPRU. This work is being updated, but the broad profile of the boiler stock is nevertheless probably valid, although the numbers will change. According to the data, British industrial boiler stock comprises a very large number of small boilers and about 420 in the range that the EEC Directive is currently addressing. These 420 are believed to be grouped in some 140

installations and perhaps 30 of them are in the 300 MW plus range.

The contribution which industrial boilers in the over 50 MW class make to the total national problem of NO_x and SO_2 is relatively small. All industrial boilers of whatever size appear to account for only about 11% of the SO_2 and 5·5% of the NO_x. Indeed, there is a view that industry has already made most of its contribution to any national 'bubble reduction'—even if it were largely a result of the recession in manufacturing industry!

Faced with these statistics many industrialists argue that any programme to reduce overall national emissions should concentrate on the really large (i.e. public utility) units. However, there is no clear national policy on this and some argue otherwise. It is worth amplifying the relevant points, taking an admittedly worst-case scenario.

Figure 5.7.1 attempts to illustrate the decision process for an industrial boiler and it must be stressed that each boiler, certainly each of the 140 or so very large installations, is different. Partly due to the capital cost involved, partly because it may be physically difficult to fit in the necessary plant, the preferred choice for most industrial boilers would be to move to a low sulphur fuel; fuel options are discussed below. Having decided to change fuel the EEC 'catch 22' yawns before us. As at present drafted, if you change fuel type you become a 'new plant'. Now the Commission's proposed new plant limits represent something like Best Available Technology for a purpose-built unit (and there are some who believe they are too ambitious even for new units). The chances of achieving these on an existing plant are remote indeed. Industry is, therefore, trying to get this aspect of the Directive changed. There must be severe doubts about how well devices like low NO_x burners will work when retrofitted to old plant and what other problems they will bring in their wake. Certainly many believe that it will be difficult to achieve anything approaching a 60% NO_x reduction.

Table 5.7.1 Industrial boiler stock: best estimate (SPRU)

1983	Below 50 MW (thermal)	Above 50 MW
Number of boilers	47 600	420 (140 units)
SO_2 (K tonnes)	307	104
% national total	8·5	2·8
NO_x (K tonnes)	70	26
% national total	3·9	1·4

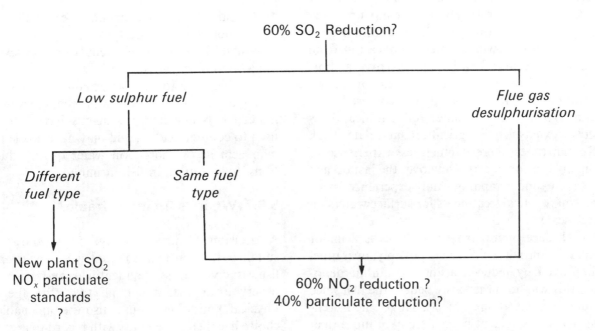

Fig. 5.7.1.

Turning back to the question of low sulphur fuel, the options look something like this.

Gas. British Gas plan to increase sales to industry by about 15%; beyond this it depends on Government policy on imports.

Oil. Wide availability depends on refinery policy and capital investment by the oil industry. This looks like the most expensive option.

Coal. There are only very limited supplies of lower sulphur coal in the UK. There is believed to be plenty of low sulphur coal worldwide, but its import would have serious implications for the UK coal industry—again a political factor. A premium would likely develop on low sulphur coal if relative demand for it grew to meet legislation.

What then does all this mean for industry? UK-based coal conversions will become much less attractive; people will be less inclined to invest in new plant because of the stringency of the new plant criteria; depending on the extent that industrial plants are required to contribute, it will increase both capital and operating costs.

There have been estimates that the Directive could increase industry's costs by about £140 million per annum by the year 2000, partly as a result of the disincentive to invest in coal conversions. Investment in monitoring equipment, discussed in an earlier section, could require in excess of £35 million and fall disproportionately on the user of smaller boilers.

Much tends to be made of the fact that any expenditure which may be stimulated by legislation on acid rain is good for the sectors supplying pollution control equipment, and this is true. However, since in practice a company's capital expenditure over a reasonable period is fixed by external factors, increased expenditure on pollution control plant, new boilers, etc., will be precisely off-set by less expenditure on production plant. At the national economic level, therefore, one sector of the equipment supply industry will be stimulated and another will be depressed. Meanwhile the operator of the boiler or other combustion plant will, all other things being equal, have his costs increased. There is therefore a cost to be borne by industry overall. Industry will also bear a part of the costs of emission reduction by the electricity generating sector, through increased tariffs. Indeed there is evidence that in West Germany electricity charges have not reflected fully the recent reduction in fossil-fuel prices because of the heavy costs of pollution abatement.

5.8 REMEDIAL STRATEGIES FOR ACIDIFICATION DAMAGE

5.8.1 Introduction

Lake or soil acidification implies a progressive loss of neutralising bases from run-off waters or catchment soils. Replenishment of these lost base materials is essential if acidified sites are to be restored and so a necessary adjunct to any proposed measures for emission control. In addition such remedial measures will counter acidification from whatever cause, they can be effective over a short time-scale and can be focussed on areas or ecosystem components where damage is recognised.

Although liming to counter soil acidification is practised extensively in agriculture, and dates back to Roman times, there is much less experience of liming uplands or forests. Only over the last decade has much understanding and experience been gained on practical treatments for surface waters on a large scale.[1]

The absence of fish from acid lakes is a major concern. While lake acidity may possibly, over time, be improved by reduced acidic or sulphur deposition, this will not necessarily benefit the fish since calcium concentrations are crucial to their well-being. This was evident from a study of more than 700 lakes in Southern Norway where those without brown trout were those low in calcium, even though of widely different acidity.[2] It is clear that if fisheries are to be restored, or established in fishless lakes, calcium concentrations must be increased.

Thousands of acid waters are being limed in Sweden, and substantial trial programmes are underway in Norway, United Kingdom and North America, providing a better knowledge of suitable methods, effective time-scales and costs for a wide variety of conditions.

Soil or catchment liming has also been promoted as a land improvement practice in many countries, especially on poor upland soils. It is usually combined with improved draining and fertiliser applications. However, it is usually a pragmatic farming procedure, and there is rather little account taken of the subsequent changes in soil composition or soil leachates or in developing a scientific understanding of the processes involved, although there have been some recent advances in this field.[3,4]

Remedies for forest decline must await identification of the cause of decline—the current view is that acid deposition can be, at most, only a contributory cause. Forests planted on poor, acid, upland soils show little response to 'lime' (i.e. calcium) applications but are often deficient in other bases (e.g. magnesium), trace minerals (e.g. potassium), or nutrients, and do respond favourably to applications of mineral or nutrient fertilisers.

Emphasis is given here to liming techniques, especially for improving acid waters: other remedial

strategies should not be forgotten, however. They include:
— land-use management, especially for moorland and forest areas;
— hydrological management, to reduce extreme flows or drying out;
— fisheries restocking and management.
In some cases a combination of techniques may be indicated. Non-calcareous agents have also been used to counter acidity, and may have a role in the management of short-term water quality fluctuations, for instance, in fish farming.

5.8.2 When is liming suitable?

A decision framework (Fig. 5.8.1) shows that suitable sites must be selected where acid and/or fishless conditions are identified and where the fishery is important enough to justify the cost. Physical conditions must also be amenable. A choice has then to be made with regard to a suitable method of application, the most appropriate material and its formulation, and on the rate of application.

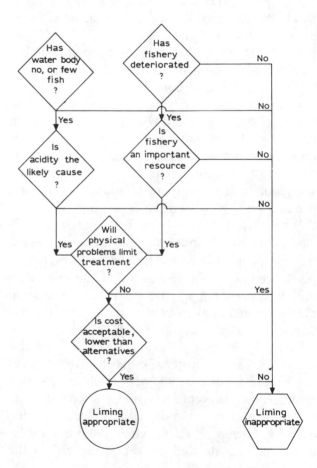

Fig. 5.8.1. Liming decision tree.

5.8.3 Target water quality

The minimum water quality which permits survival of a fish population depends on the sensitivity of the most vulnerable stage in the life history. For brown trout, for example, the freshly fertilised egg is sensitive to low pH and low calcium but is relatively insensitive to aluminium. As it develops the eyed egg becomes relatively insensitive to water quality, but then becomes very sensitive to acid, low calcium and aluminium as it emerges from the egg membrane.[5] The yearling fish develops more tolerance to acid and low calcium, but becomes progressively more sensitive to aluminium (Fig. 5.8.2). There are, moreover, interactions between the toxicity of these chemical components (Fig. 5.8.3) so that the toxicity of low pH can be offset by high calcium concentrations and vice versa, and high concentrations of aluminium exhibit their peak toxicity around pH 5 with low calcium concentrations (Brown, 1983),[6] possibly because of a greater proportion of the total being in the toxic form. For brown trout we can define a minimum threshold for calcium and maximum thresholds for acid and aluminium; water quality of pH > 5, Ca > 2 mg litre^{-1} and Al (total) < 100 µg litre^{-1} should protect all life stages from acute effects.[7]

However, target water quality should never be set at the level of acute toxicity, and to ensure safety both during episodes, as well as to provide conditions suitable for growth and reproduction, average water quality should be at some defined level above the observed effect thresholds. It is important to note that since different life stages have different sensitivity and may inhabit different parts of a river system, a liming programme should identify the life stage or location to be protected. Since adult fish are less sensitive, it follows that protection of a put-and-take fishery is less demanding than for a self-sustaining population.

Most research has been done with salmonid species which are more sensitive to acidity than most other freshwater species.[8] In northern Europe, upland waters sensitive to acidification are usually salmonid, rather than 'coarse fish' waters. In Sweden, however, many acid lakes have mixed species populations and roach and perch are reported to be affected. Much less is known about the specific water quality requirements of these species but they are likely (for basic physiological reasons) to be similarly dependent on a minimum calcium level and on maximum levels of acidity and aluminium, although possibly with different thresholds.

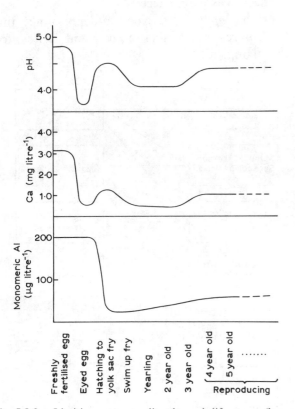

Fig. 5.8.2. Limiting water quality through life stages (brown trout).

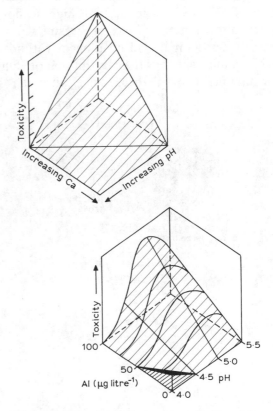

Fig. 5.8.3. Toxicity inter-relationships of pH, calcium and aluminium.

5.8.4 Application sites

Sources and pathways of acidity in the ecosystem are various, and they are illustrated in Fig. 5.8.4. They include

— atmospheric deposition
— catchment soils and vegetation
— tributary streams
— the lake water body
— lake sediments.

Leaving aside the first (via emission control or atmospheric base dispersion), base materials can be applied to the catchment and its vegetation, to the tributary streams, to the water body, or to lake sediments. Part or all of these components can be treated, for instance, by limiting lime applications to only part of a catchment, or to headwater bogs and wetlands, or to riparian zones of streams or lakes. Choice of the preferred option must vary with the objective and with site conditions, as well as on the balance to be struck between the benefits, and any disadvantages and costs. While adding lime directly to a water body is relatively cheap and immediately effective, its benefit is limited to the time-scale of hydrological replacement, and it will have to be repeated to maintain a defined target water quality. It is more difficult to lime streams directly since need is greatest at highest flows. Further, direct liming to lakes or streams will not reduce the substantial acid component within the soils of acidified sites nor the contribution of aluminium mobilised from these soils, although its

toxicity may be moderated by the raised calcium levels. Liming soils of a catchment has the benefit that the treatment is longer lasting (i.e. over years), and that once the base reserve of the soils is restored, aluminium should be retained. The extent to which lime is retained in the soils, rather than being released over time, is still not certain for many of the acid soils treated. It is also likely to be more expensive and benefits to surface waters may not be immediately evident if the soils are below their water capacity.

5.8.5 Application methods

Preferred methods of application are dependent on site conditions and the local availability and costs of lime materials, as well as the accessibility of target areas. The options are summarised in Fig. 5.8.5. They include:

— *aerial applications* using hoppers, sprays, or 'bomb' techniques;
— *truck or tractor applications*; these are suitable for distribution of quite large quantities from a local depot, but need reasonable road/truck access;
— *manual or tractor applications*; these are suitable for small-scale or pilot projects and need reasonable access;
— *slurry application* from a depot tank; this requires reasonable access and water for mixing;

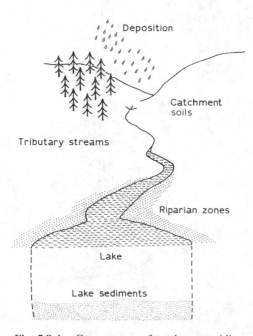

Fig. 5.8.4. Components of catchment acidity.

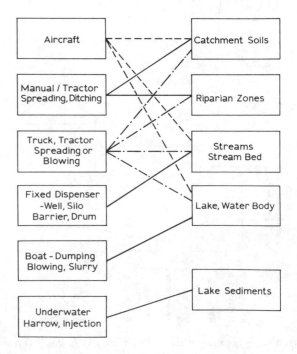

Fig. 5.8.5. Choice of application method.

— *boat application*; simple and relatively cheap but needs a depot of material on shore; can be applied dry or as a slurry; it is slow for large applications and requires access by road or water.

Fixed dispensers allow more precise dosing in response to need, activated by stream flow or pH. Installations also require capital initially and then resources for inspection and maintenance; they may not prove suitable in all flow conditions and are not suitable for remote sites.

Liming materials or formulations are selected on the basis of their yield of calcium or base, but some effective neutralisers are difficult to handle or have other disadvantages such as releasing humic materials. Chalk or limestone (depending on local sources) are the usual materials, but geological origin and the presence of magnesium (ineffective for fish), or of potentially toxic trace metals, will influence choice. Carboniferous limestones and Cretaceous chalks are more soluble than Permian or Jurassic material. Particle size is a particularly important determinant of calcium availability over time—generally the finest size spectrum will be more effective since solubility is governed by surface area, but it is more difficult to handle and may be more expensive.

Possible liming materials include:
— *calcium oxide/quicklime*: rapidly soluble and highly reactive but its effect is short-lived and handling is hazardous; application to soils may be unsightly at high loading rates;
— *calcium hydroxide/slaked lime*: rapidly soluble and highly reactive, less caustic than calcium oxide but still difficult to handle and unsightly at high application rates; effective only over a short term;
— *pulverised limestone*: freely available as an agricultural material, solubility and reactivity

less than for lime, but safe to use and effective for a longer period;
— *limestone rock*: may be freely available and easy to handle but solubility and reactivity are slow;
— *shells/shell sand*: available in limited quantities and at specific locations; soluble and quickly effective even when not finely particulate.

These qualities are summarised in Table 5.8.1.

Sodium salts and even sea water have been used in experimental projects.[9]

5.8.6 Cost, cost-effectiveness

Significant factors in the cost of liming applications are the size and location of the site, the quality of the bedrock, soils, sediments and run-off water, the target water quality, the acid 'loading' from atmosphere, the type of vegetation and its management, and the hydrological turnover. Cost estimates must take account of project planning, initial site characterisation, and the source and characteristics of material to be used. Fixed installations or equipment to apply the treatment material will incur capital costs. Monitoring for the expected water quality improvement, fishery re-establishment, and reapplications or restocking will all need continued resource allocation.

Cost does not equate with cost-effectiveness, however: the latter is dependent also on the volume of water or area to be treated, and on the time-scale of hydrological turnover. Thus, for streams, volume is the product of flow rate and time. Cost is also dependent on the pretreatment water quality and the extent to which the water quality is to be changed. Although fish survival/restoration is the ultimate aim, it is simpler to define the target in terms of water quality, where alkalinity represents the capacity to buffer acidity (and also in this case

Table 5.8.1 Neutralising agents compared

Agent	solubility	Overdose capability	Corrosivity	Duration of effect	Neutralising value[a]	Cost[b]
Limestone rock	++	(0)	(0)	++++	<100%	Low
Agricultural limestone	+++	(0)	(0)	+++	<100%	Low
Chalk	+++	(0)	(0)	+++	<100%	Med.
Olivine, $(MgO)_2SiO_2$	+	(0)	(0)	++++	~50%	Low
Calcium oxide	++++	++++	+++	+	180%	High
Calcium hydroxide	++++	+++	++	+	136%	High
NaOH	++++	++++	++++	+	250%	High
Basic slag $(CaO)_2SiO_2$	+	(0)	(+)	++++	~50%	Low
Shell, shell sand	++	(0)	(0)	++	<100%	Med.

[a] Relative to $CaCO_3$.
[b] Takes account of availability and transport.

provides a measure of available calcium). Cost-effectiveness (CE) is thus:

$$CE = \text{total costs/volume} \times \text{alkalinity}.$$

This expression yields a value in terms of unit volume and unit change in alkalinity and so facilitates comparison between treatments or between projects. A Swedish study of 30 projects (Fig. 5.8.6) shows that boat applications were always cheapest, trucks dearest, and helicopters most versatile in Swedish conditions. Land applications were considered less effective in the short term but were appropriate for longer lasting action.

Costs should include those incurred in re-establishing the fishery, including restocking adults and fry, establishing hatcheries, and possibly installing refuges for worst conditions if treatment does not prevent surges of acidity.

5.8.7 Potential problems

A number of potential problems have been identified.

Lakes or streams which are most susceptible to acidification are also those with rapid hydrological turnover and high flows and so will be those most difficult to treat. They may also be those which suffer most from climatic extremes, e.g. frost and drought, and these effects must be distinguished from those attributable to acidity.

If aluminium is the limiting component for fish survival, lime must be applied to the catchment not the water body. As aluminium is most toxic at pH 5·2 to 5·5, conditions immediately after liming directly to a lake may not be suitable for fish. This can be avoided by monitoring for the change in pH but delaying restocking until the water quality is satisfactory.

Liming may precipitate phosphate and so impoverish waters still further; however, evidence for this is scarce and not altogether clear. Most acid waters are naturally low in phosphate and as a consequence they will have a low potential fish yield.

Lime applications at suitable loadings may alter the vegetation and consequently its dependent fauna. Selective liming of only part of the catchment could limit effects of this kind if they are considered undesirable.

Repeated treatments result in periodic changes and non-equilibrium conditions; an effective schedule of reliming can be designed once the hydrology and lake response is understood.

5.8.8 Experience

Notwithstanding these potential difficulties, there is a record of successful treatment of many lakes and streams. As experience and understanding develop, effective liming programmes can be designed and carried out to minimise problems yet still achieve their objective. In Sweden more than 4000 lakes and streams are being treated with Government funding (Skr 70 million in 1984/85). Fisheries have been restored and demand is so strong that it is difficult to retain acid lakes for scientific study. In Norway 177 lakes and streams in the south have been treated with Government support (Nkr 4 million in 1984/85).[9] In North America many substantial research investigations are underway,[10,11] and a major liming programme is in hand to improve Adirondack lakes.[12] In the United Kingdom several trials are in operation with Government and industry funding.[13-15] The sites include lakes in west Wales and south west Scotland and catchment treatments in south-west Wales and south-west and central Scotland. These treatments include land-use changes such as forest cutback, or replanting with broadleaf species, upland improvement techniques, and burning and ploughing regimes. Early results are favourable in terms of water quality.

5.8.9 Conclusions

Liming can be demonstrated to be a practical, economic and successful strategy to overcome acid water conditions. Although Sweden's 4000 limed lakes are not routinely monitored for fish, some

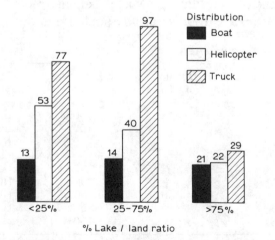

Fig. 5.8.6. Relative treatment costs, based on 30 Swedish projects.

studied lakes have shown both natural recolonisation and successful restocking of important fish species, including salmon, trout, char and perch. The Swedish Government's current investment in liming is an index of that success. In contrast, the recent reduction in sulphur deposition in Sweden has not led to an improvement in the pH of Lake Gardsjon although liming has done so.[16] Nor has the experimental exclusion of acidic deposition in Norway provided the expected pH change over the period of treatment (more than 2 years).

The costs of emission control, even for the United Kingdom alone, are estimated to be substantial. Liming all Sweden's acid lakes would cost less than emission control at a single 2 000 MW fossil-fuelled station. The current view is that deposition at acidified sites would have to be reduced by more than 50% (EEC, 1986),[17] yet many of these sites have about 50% deposition attributable only to 'background' so that industrial and other emissions from all of Europe (not just the Community) would have to be zero. In view of the political, practical and scientific uncertainties of such a strategy, liming or similar local strategies offer a more certain and immediate success.

REFERENCES

1. FRASER, J. E. & BRITT, D. L., Liming of acidified waters: a review of methods and effects on aquatic ecosystems. *U.S. Fish & Wildlife Service*, FWS/OBS-80/40.13 Washington DC, 1982.
2. WRIGHT, R. F. & SNEKVIK, E., Acid precipitation: chemistry and fish populations in 700 lakes in southernmost Norway. *Verh. Int. Verein. Limnol.*, **20** (1979) 765–75.
3. WARFVINGE, P., Neutralization of Soil Systems. Licentiate Thesis, University of Lund, Sweden LUTKDT/TKKT/1002/1-220, 1986.
4. WARFVINGE, P. & SVERDRUP, H., Soil liming and runoff acidification mitigation. *Int. Symp.: Lake and Reservoir Management, Oct 18–20, Knoxville, Tenn.* EPA 440/5/84-001, 1984, 389–93.
5. BROWN, D. J. A. & LYNAM, S., The influence of calcium on the survival of eggs of brown trout (*Salmo trutta*) at pH 4·5. *J. Fish Biol.*, **19** (1982) 205–11.
6. BROWN, D. J. A., The effect of calcium and aluminium concentrations on the survival of brown trout at low pH. *Bull. Environ. Contam. Toxicol.*, **30** (1983) 582–7.
7. HOWELLS, G., BROWN, D. J. A. & SADLER, K., Effects of acidity, calcium and aluminium on fish survival and productivity. *J. Sci. Food Agric.*, **34** (1983) 559–70.
8. HOWELLS, G., Acid waters: effects on aquatic systems. *Adv. Appl. Biol.*, **9** (1983) 143–255.
9. ROSSELAND, B. & SKOGHEIM, O., Attempts to reduce effects of acidification in fishes in Norway by different mitigation techniques. *Fisheries* **9** (1984) 10–16.
10. SCHREIBER, R. K. & RAGO, P., The Federal plan for mitigation of acid precipitation effects in the United States: Opportunities for basic and applied research. *Fisheries*, **9** (1984) 31–5.
11. SCHEIDER, W. & BRYDGES, T. G., Whole lake neutralisation experiments in Ontario: a review. *Fisheries*, **9** (1984) 17–18.
12. KRETSER, W. & COLQUHOUN, J., Treatment of New York's Adirondack lakes by liming. *Fisheries*, **9** (1984) 36–41.
13. Welsh Water Authority, *Acid waters in Wales*, 1986.
14. BURNS, J. C., COY, J. S., TERVET, D. J., HARRIMAN, R., MORRISON, B. R. S. & QUINE, C. P., The Loch Dee project: a study of the ecological effects of acid precipitation and forest management on an upland catchment in south west Scotland. *Fish. Management*, **15** (1984) 145–68.
15. HOWELLS, G., Acidity mitigation of a small upland lake. In *Proc. EEC/COST Workshop, Grimstad, Norway 9–11 June 1986.*
16. HULTBERG, H., Changes in fish populations and water chemistry in Lake Gardsjon and neighbouring lakes during the last century: In *Lake Gardsjon: an acid forest lake and its catchment*, ed. F. Anderson and B. Olsen, *Ecological Bulletins*, No. 37 (1985) 64–72.
17. EEC, *European Community/COST Workshop on Reversibility of Acidification*, Grimstad, Norway 9–11 June 1986.

5.9 DESIGNING EFFECTIVE CONTROL AND ABATEMENT POLICIES

5.9.1 Introduction

Amongst the many uncertainties in what may broadly be called the acid rain debate, there is one fact that is quite clear: achieving large reductions in potentially polluting emissions is an expensive business. For the UK the requirements of the proposed European Directive, for example, would be likely to cost many hundreds of millions of pounds annually. This is not to say that such expenditure cannot be justified. Rather it argues for making the maximum use of what is known, or what seems reasonably likely, in drawing up policy and legislation concerning atmospheric emissions. The need to establish more of the facts of the situation so as to aid policy is well rehearsed, and is accepted by the UK government. There are now many national and international efforts to understand more of the science of acid rain phenomena. However, the question still remains as to how the current and anticipated state of knowledge can be used constructively to support policy. This question too is one for which science—of the systems and economic science variety—has a role to play. The broad purpose of this paper is to review the methodologies and considerations that are relevant in helping design effective policies. Quite what is meant by 'effective' is discussed later. However, amongst other things it includes prominently the need to achieve policy objectives at minimum cost.

5.9.2 Cost-benefit analysis

With almost any significant policy decisions costs will be involved as well as hoped-for benefits, and there may be winners and losers amongst the different groups involved. Cost-benefit analysis (cba) is a framework in which all of the implications of a decision are identified, and are then converted to the common currency of money so that the nett result of the decision (with all its pluses and minuses) can be evaluated. Decisions shown to have a nett positive value are then worthwhile in the sense that they make society richer. There are many instances when cba is useful, but these will generally be where it is fairly straightforward to identify all of the consequences of a decision and to express these in monetary terms. Often this is difficult and although ingenious methods have been designed to evaluate effects, those valuations are sometimes unconvincing. It has been widely used in the environmental protection area. For example, it is used by the UK Industrial Air Pollution Inspectorate and is required by law for some aspects of US air pollution administration.

The approach has been applied to evaluating international pollution abatement proposals but here the ground is much weaker and there are many reservations that can be made both about the appropriateness of the method, and the validity of the data/assumptions. The magnitude of the task is illustrated in Fig. 5.9.1, which shows the main stages involved. The implementation of the policy will be through control technologies, the deployment of which requires expenditure (costs of control) to achieve emission reduction. The reduction in emissions next has to be translated into a change in deposition of critical chemical species, over a European (where that is the area of interest) grid. For each element of the grid it is necessary to estimate the stock of each sensitive item at risk, and to estimate any resulting damage reduction using a model relating damage to exposure. Finally, it is necessary to translate these estimates of reduced damage to the stocks at risk into money terms, so that they may be directly compared against the costs of control.

Despite the formidable difficulties, an OECD study reported in 1981 did apply cba to estimate the value of reducing SO_2 emissions in Europe. Although the declared purpose of the report was just to illustrate a methodology, results were quoted suggesting that some measure of reduction would be cost-effective. Perhaps unsurprisingly, this result was quite widely reported without the qualifications made in the report. The study made many assumptions that were highly controversial at the time and, just a few years on, the consensus would be that the numbers arrived at are of quite limited value. For example, the improvements to human health and life expectancy were suggested as potentially the most major benefit of reduced emissions, whereas it now seems quite probable that current SO_2 levels

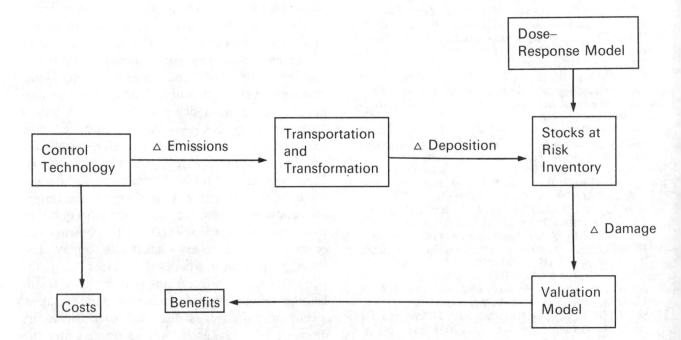

Fig. 5.9.1. Simplified scheme for cost-benefit analysis of emission reduction.

cause no, or only minimal, health effects. Neither was the role of pollutants other than SO_2 considered. More recently, in introducing its proposed Directive on Large Plant Emissions, the European Commission has quoted results suggesting that the benefits would be greater than the costs involved. Notwithstanding the difficulties on the benefits estimation, this analysis also appears very seriously to underestimate the costs of the policy, by neglecting the need for retrofitting control technologies. Nevertheless, the argument that 'emission reduction is free', in the sense that expected benefits exceed costs, continues to be used by the Commission.

These examples suggest that an over-ambitious use of cba, going beyond the limits of the data, can mislead rather than aid policy. In addition to problems of data collection, there are also theoretical reservations about using cba in such circumstances. These relate primarily to the problems of adequately valuing many of the environmental benefits objectively and in money terms. For example, there may be no objective middle road between, on the one hand, how a German values his forests or a Swede his lakes, and on the other, the price of timber or fish in the local market. Such valuations depend fundamentally on who the decision-maker is.

The true cost, as opposed to direct money cost, of a policy is also not simple to define where capital is limited. Where expenditure comes from the public purse the cost is, perhaps, benefits foregone from increased expenditure on education or health. Private sector expenditure on environment might force out expenditure on other wealth-producing activity. Conversely, tough legislation might of itself produce a highly productive control technology industry, throwing off secondary economic benefits. Thus neither side of the cost-benefit equation may be quite what it seems, and convincing demonstrations that one or other emission reduction policy 'pays for itself' are likely to be elusive.

5.9.3 Decision support models

Frameworks are being developed which, like cba, bring together a comprehensive account of the costs and consequences of pollution control policy but, instead of seeking to provide a single numerical cost answer, are designed so that decision-makers can explore alternative policies and can apply their own value system to each set of consequences. Compared to the cba framework of Fig. 5.9.1, these models avoid the monetary evaluation stage. Instead the outputs are multi-dimensional statements about damage reductions. Developments with many points in common are being undertaken at Cambridge University, and at the Vienna-based International Institute for Applied Systems Analysis (IIASA). Both models aim to help decision-making within the UNECE. The Cambridge University work is being supported by the UK Department of the Environment. It is fairly comprehensive in the damage categories considered, but the damage mechanisms are represented by simple dose-response functions. To represent the great uncertainty in damage mechanisms, the model user can choose from a range of these functions. Only sulphur dioxide mechanisms are explicitly included, but the researchers consider that the effects of other pollutants can be rolled into the general uncertainty of sulphur dioxide effects. The IIASA model RAINS (Regional Acidification INformation and Simulation) at the outset was much more limited in the damage categories considered. However, it includes much more detail of the scientific understanding of the mechanism involved. The model is becoming more comprehensive and nitrogen oxide emissions will be explicitly included. This work is able to draw upon an international network of collaborating institutions.

Undoubtedly the IIASA model is the better repository of the state of scientific knowledge, but the Cambridge work seems to have the more sophisticated approach to uncertainty. It may be that this is more important in the present state of play. At the minimum such models should help those responsible for policy to take a comprehensive look at the issues and to get a better feel for the magnitude of the effects. Both utilise modern computer technology to allow users conveniently to explore the implications of many alternative policies. Current work by the Beijer Institute, at York University, utilises a somewhat different theoretical approach and seeks to identify the minimum-cost ways of achieving stated environmental objectives. The model covers Europe and the environmental objectives are to ensure that acid deposition in sensitive areas does not exceed target levels. These levels are agreed through discussion with experts prior to a run and, of course, the consequences of changing the targets can be explored. By adopting this approach the researchers have avoided the need to include dose-response relationships or valuation measures in their model.

In contrast to the work described above, this

model uses linear programming methods to derive the minimum cost strategy, rather than evaluate the consequences of some strategy fed into the model. It might be expected that the cost optimum strategy will not be one which imposes uniform requirements across the countries considered; instead the model is able to take account of the differing opportunities for emission reductions amongst European countries, and differing proximities to sensitive areas. The general form of the model is quite flexible so that it should in principle be possible to incorporate additional constraints to reflect what is likely to be politically acceptable.

5.9.4 Minimum-cost emission reduction

The UK, amongst other nations, is under pressure to reduce emissions irrespective of the conclusions that may come out of the trans-European system models described above. It is, therefore, useful to address the much more limited question of how an individual state can derive a policy for achieving target emission reductions at minimum cost.

The theoretical basis for minimum-cost emission reduction is the common sense approach of using the cheapest means first. In essence there are two stages: first, one must assemble an inventory of the alternative means of reducing emissions ('fixes') with the cost of each in terms of £ per tonne of pollutant removed. Second one must determine the scope for applying each in the setting considered, for example, the UK. The 'fixes' are then applied in order of cost until the desired level of emission reduction is reached. The 'fixes' may be emission-reducing technologies, such as FGD, or they may represent fuel switches or conservation. In principle, a 'fix' may have a negative cost, as would be the case, for example, if a lower sulphur fuel had a lower cost than some fuel currently used. Switches of this kind should occur through the normal operation of the market irrespective of environmental policy. Major examples are the penetration of natural gas in the UK, and nuclear power in France.

Over the past few years the costs and capabilities of many emission-reducing technologies have become established and there is active international information exchange, for example, through the many conferences on this topic that take place. Within OECD there is a formal data collection exercise to establish costs, which is at present focusing upon ozone precursors.

The lowest cost route to emission reduction will depend upon the particular characteristics of the society under consideration. It will need to take account of the existing energy supply pattern and the costs and availabilities of indigenous fuels. Also factors such as the scope for cost-effective energy savings will differ between settings. We must therefore expect many differences amongst the most cost-effective routes for individual countries. There have been more or less informal exercises to establish the best means of achieving reductions, along these lines, in several countries including the UK. A more formal exercise has been reported for a region of Germany.

A rigorous approach to establishing a minimum-cost policy is more complex than that described. It would be necessary to address energy demands as they are likely to evolve in the future, and changes in energy use will themselves be influenced by the environmental legislation selected. Also individual technologies may be capable of different levels of control but at different costs; in general the more emissions are reduced the higher the cost. Such considerations make the detailed derivation of the best approach for a nation quite difficult to establish. However, the key elements of a policy should be quite straightforward to identify.

5.9.5 Legislative mechanisms for emission reduction

The ultimate objective of environmental policy regarding atmospheric pollution is to secure the conditions under which any future environmental damage is minimised, and past damage is reversed. In the main the expression of this objective is through legal mechanisms which control and reduce emissions. The mechanisms available are:

(i) emission limits on plants. These may relate to all plants, or plants only over a certain size, or just new plants. Emission limits are usually implemented through issuing operating licences only to conforming plant;

(ii) 'bubble reductions', requiring overall emissions from some stated geographic unit not to exceed particular limits, or to reduce by some stated fraction. 'Bubbles' have been discussed for units of very different sizes. At one extreme, the EEC has been thought of as the bubble. More usually, individual nations form the bubble unit, or in the USA, industrial estates or individual factories;

(iii) control of fuel used. In the main this addresses the concern to reduce sulphur oxide emissions; or

(iv) approaches based on the standard of performance required, rather than the standard of emission level. These include, for example, a requirement to use 'best available control technology' or 'as low as reasonably achievable'. The discussion here will be limited to the familiar 'best practical means' approach which combines technical and economic considerations.

In addressing the advantages and disadvantages of these mechanisms, it is useful to map out some characteristics of what was described earlier as an 'effective' policy. These include:

(1) low cost, that is, the environmental objective should be achieved at the lowest feasible cost;
(2) general applicability across different national conditions;
(3) easy to police;
(4) direct link with the state of technology so that the requirements can be updated as control technology improves. Legal requirements might also be designed to force improvements in the available technology;
(5) easy to integrate with other policy objectives, for example, on industrial development or energy policy;
(6) capable of achieving high environmental standards; and
(7) perceived equity.

Other attributes might be added to this list, and certainly there is scope for disagreement on the relative weightings that might be given to each. Nevertheless, these seven headings form a useful checklist against which the available legislative mechanisms might be considered.

5.9.5.1 Emission limits

Internationally, plant emission limits have been the main means for securing emission reductions. Usually restrictions have been placed upon the larger new plants, but regulation has been extended, for example, in Germany, to existing plant with more than a specified life expectancy, and down to smaller plant.

An advantage of plant emission standards is that they can be directly linked to technology, so that standards can be set for particular types of plant which fully exploit technological capabilities. Thus

'excessive' costs can be avoided, although what is considered excessive is essentially a political question. Any particular technology can, of course, achieve reductions in emissions only up to a limit. If the legislation pushes beyond that limit, then an expensive switch in technology could be required. The trick in setting emission standards is to find a value which gets the most out of a relatively cheap technology without forcing some users to switch to a more expensive one. Where plant design, energy demand patterns, and fuel used are uniform, then it is reasonably easy to set a standard. However, on a European scale many differences exist and it is not obvious that any simple set of common standards could be effective. The problem can be illustrated by the case of nitrogen oxide reductions. Combustion modifications can reduce nitrogen oxide emissions in power-generating plant, and low-NO_x combustion is a quite cost-effective approach, being very much cheaper than the alternative of catalytic reduction of the nitrogen oxides in the flue gas. However, the extent of reductions achievable by combustion modification depends, to an important extent, on the plant design. If standards are set according to what is achievable by the most 'difficult' plant, then operators of other plant have no incentive to utilise fully the scope for NO_x reduction, and so a potentially cost effective means of reduction will be lost. If the standard is geared to the 'easier' plant, operators of difficult plant will be forced to adopt more expensive alternative measures.

In principle, the legislation could define alternative standards for each type of plant. However, on a European scale, there would be very many types of plant if all the relevant differences were to be reflected and legislation would be unworkably cumbersome.

Addressing some of the other policy attributes, emission standards are expensive to police when, as in the case of the European proposals, they require monitoring at each plant. They also allow rather limited flexibility to integrate with other policy objectives. Within the European Community it has been argued that emission limits are fair because they impose explicitly similar requirements on operators of similar plant. Indeed, it is claimed that Community-wide emission limits are necessary to avoid distortions of competition. However, since different members of the Community differ in the extent to which they would be affected (for example, through varying energy intensity, or use of nuclear power), the overall influence on competition is not clear.

5.9.5.2 Bubble reduction

Bubble reductions require some defined geographic unit to decrease its emissions by a specified amount, generally over a specified period of time. In the European context, the bubble reductions most discussed are those applying to individual nations. Most usually they are stated as a requirement to reduce future emissions to some stated fraction of 1980 emissions. The great advantage of bubbles is that they allow the means of reduction to be selected entirely at the discretion of those within the bubble. In principle, therefore, countries could use some mix of measures either to approach as closely as possible the theoretical minimum cost solution, or could use the flexibility to achieve coordination with other policy objectives, for example, using indigenous fuel supplies.

Although the freedom allowed within the bubble gives the possibility for cost-effective reductions, the reference year for base emissions, the target year for reduced emissions, and the percentage reductions required, are all essentially arbitrary and can make a great deal of difference to the costs involved. The selection of 1980 as a base year does not suit the UK because no benefit is derived from the great emission reductions made over the previous decade. Similarly a mid-1990s target year would not anticipate any benefits from increased penetration of nuclear power. For these reasons uniform requirements across all countries of Europe may not necessarily be perceived as equitable. They would certainly cause problems for countries currently having relatively low energy use and emissions, but intending to increase energy consumptions.

Policing bubble reductions might be less onerous than emission limits since it may not be necessary to monitor all plants individually; it is possible that some sampling method could be agreed. However, it is necessary to have a suitably validated account of emissions for the base year and this is not available for some European countries for 1980.

The link between bubble reductions and control technologies is not direct and it can be difficult to assess how improving technology and lowering costs should be reflected in the bubble requirements.

In the USA, bubbles on a much smaller scale are popular. Rather than using central edicts on how reductions are to be achieved, market mechanisms are used so that, for example, companies within the same bubble might trade emission allowances. One company might undertake greater reductions than required, at a price, in order that a second company could avoid any need for action.

5.9.5.3 Fuel restrictions

Sulphur oxide emissions could be reduced by limiting the allowable sulphur levels in coal and oil. This is likely to be the only practical way to reduce emissions from the smallest of users. The additional cost of low sulphur oil and coal against average sulphur oil and coal depends upon the supply/demand balance for low sulphur fuels. The premium on internationally traded low sulphur coal is not large at present but would be expected to increase should new legislation increase demand. Quite large amounts of relatively low sulphur (say 0·8%) coal could become available on world markets, but is not produced in the UK. This coal would contain around half the average of UK-produced coals. The potential supply position of low sulphur fuel and heating oils is much tighter. Technology exists for removing sulphur from fuel oil, but estimates by the oil-industry group CONCAWE suggest this would not be cost-effective. Implementation of fuel restrictions might be seen to favour those countries for which it involved least disruption of existing supplies.

5.9.5.4 Best practical means

This is an essentially pragmatic system used in the UK to ensure that a satisfactory balance is maintained between the potential environmental damage, and the cost of control and ability to pay. What 'best practical means' implies for different classes of plant can be specified but presumably the requirement might be varied to take account of individual circumstances. It is possible to link best practical means with technical development. The flexibility inherent in the system should allow cost-efficient emission reduction. The great difficulty with the approach is ensuring that the required aggregate reduction in emissions will be achieved. Were the approach to be adopted on a European scale there would doubtless be difficulties in ensuring a uniform interpretation, and the application would reflect the individual political stances of the different countries. Thus there might be doubts about its fairness.

Table 5.9.1 summarises the comments made above. The immediate conclusion is that the available mechanisms have different strengths and weak-

Table 5.9.1 Characteristics of approaches to emission reduction

	Plant emission limits	Bubble reductions	Restrictions on fuel use	Best practicable means
Relatively low cost	Generally not	Yes	Future premiums on compliance fuels unknown	Yes
Uniform international application	Restricted —depends on boiler stock & fuel use	Problems for small countries	Limited scope	Problems
Link with technical advances	Good—direct	Not direct	No	Yes
Integration with other policy objectives	No	Yes	No	Yes
Easy to police	Expensive when individual monitoring required	Some problems —base inventories	Yes	Yes
Can achieve high environmental standards	Yes	Yes	No	Untested
Perceived fairness	Problems —technology applicability	Problems —start and end years	Problems —fuel availability and use	Problems —uniform applicability

nesses, and no approach is likely to meet all the criteria posed. In particular, no approach is likely to be perceived as being fair to all the countries involved. The resolution of what approach finally to adopt will be a political decision which trades off the various attributes.

5.9.6 Assessing effectiveness

Implementing a programme of European-wide emission reductions is, to a large degree, an act of faith. Scientific understanding is not sufficient to predict likely effects with any degree of confidence. It is important, therefore, to utilise this as an opportunity for a large-scale experiment so that in ten or fifteen years time we are in a stronger position. The changes in emissions seen over the past ten or so years have aided understanding only to a disappointing extent. However, these were in most cases unplanned, and there were conflicting trends in different parts of Europe, and for different pollutants. Scientific hypotheses were also much less advanced

a decade ago and monitoring was not extensive. All of these factors will have changed for the better in relation to a prospective European-scale programme. However, the experience of the last decade shows that clarification of key scientific issues does not come about just through varying levels of emissions; monitoring of effects needs to be planned.

A useful starting point would be a review of the various and sometimes competing hypotheses associated with acid rain phenomena together with an assessment of which issues might be resolved by observations over, say, a decade with 30% reductions in sulphur and nitrogen oxide emissions. This should establish what monitoring would be necessary. A programme of monitoring might then be set up and funded at the same time as embarking upon the programme of emission control. Proceeding in this way would provide a useful injection of scientific discipline into the debate by allowing prior hypotheses to be tested, rather than theorising by sifting through the entrails of past experience.

Attempts might also be made to monitor systematically the costs of implementing emission control. This is much more than the trivial exercise of counting the number of flue gas desulphurisation (for example) installations and multiplying by the unit cost. Emission control could have important secondary economic effects, both positive and negative. Control costs might sound the death knell for some industry; alternatively, for others control technology might provide new commercial opportunities. Control might give a welcomed push for more efficient use of energy, and almost certainly will shift the pattern of energy demand and relative fuel prices. In the current debate there are major differences of view on what the real economic costs might be. Clearly a proper account of these costs based upon careful analysis of experience will be very useful in deciding extensions of acid rain policy.

Finally, there should be some explicit statement of the objectives expected to be achieved. Without this it will be quite impossible to say at any stage whether the policy is succeeding or not. The objectives need, of course, to relate to the protection of environmental resources. Emission reductions are only a means to that end, and not a valid end in themselves. Monitoring will need to be set up to assess the achievement of objectives. It is clear that policy objectives would not be static; more extensive and more demanding standards of environmental protection will be desired over time. However, being clear about objectives should allow the policy debate to be more focussed, and allow more flexibility in the types of policy considered. If some small percentage of the cost of monitoring boiler emissions were earmarked for monitoring the environment, we would be better served.

5.9.7 Summary and recommendations

(i) The complexity of the scientific and policy issues puts the establishment of a 'best' environmental policy for the UK or for Europe quite beyond the reach of current methods. Nor is any proposed policy likely to be convincingly demonstrated as cost-effective, in the sense that demonstrated economic benefits exceed costs. Abatement policy is, therefore, going to have to depend heavily on political and scientific judgements.

(ii) Several models under development should aid policy formulation by allowing the consequences of alternative judgements to be explored. These models can be useful in working towards consensus both on objectives and means of environmental policy.

(iii) The costs and some consequences of a programme of emission abatement can be greatly dependent on the means of implementing the reductions. In particular, bubble requirements allowing flexibility to individual countries have potentially much lower costs than community-wide plant emission limits. There are other associated advantages and disadvantages. For the UK work should continue towards establishing the preferred means of meeting Community or pan-European obligations.

(iv) At the same time as agreeing programmes of emission abatement, monitoring regimes should be set up to ensure that maximum scientific understanding is being derived from the reductions in emissions, that environmental objectives are being achieved, and that the true costs of abatement, with associated positive and negative secondary economic effects, are established.

Appendix 1

Nineteenth Consultative Conference of The Watt Committee on Energy

AIR POLLUTION, ACID RAIN AND THE ENVIRONMENT

On 4 December 1985, The Watt Committee on Energy held the nineteenth in its series of conferences — usually described previously as 'Consultative Council Meetings'. The theme was 'Air Pollution, Acid Rain and the Environment', and the conference was held at the Institution of Mechanical Engineers. Those present were the secretaries and appointed representatives of the member institutions of the Watt Committee and others with professional interests in the subject of the Conference.

The programme of the Conference, which is printed below, included informal presentations on behalf of the sub-groups into which the working group had divided itself. The speakers included the chairmen of the sub-groups on the formation and distribution of air pollution and on damage to soils and vegetation, but not that on freshwater. The greater part of the day was devoted to the work on buildings and other non-living materials, which had not been dealt with in the Watt Committee's previous Report on Acid Rain, and to control and remedial strategies. There were several periods of discussion.

The papers presented at the Conference were early versions of those published in the Report; later, much additional work was carried out by the working group before the papers were felt to satisfy its objectives. Points that arose in discussion at the meeting were considered by the working group and taken into account in the preparation of the final version of the papers.

As Chairman of the working group, Professor Kenneth Mellanby has since summarised its conclusions in the introduction (p. ix).

Programme

Official Opening by Dr J. H. Chesters, OBE, FRS, FEng, Chairman of The Watt Committee on Energy

Session 1 Chairman: Sir John Mason, CB, FRS (Royal Society)

Introduction: Understanding 'Acid Rain' and Looking for Solutions
Prof. K. Mellanby (Chairman, Watt Committee Working Group on Acid Rain; Institute of Biology)

Formation and Distribution of Air Pollution in the United Kingdom
Dr F. B. Smith (Meteorological Office, Bracknell; Royal Meteorological Society)

Damage to Soils and Vegetation: Remaining Gaps in Knowledge
Dr W. O. Binns (Forestry Commission, Farnham; Institute of Chartered Foresters)

Pollution Processes and Recent Trends in Air Quality
Dr M. L. Williams (Warren Spring Laboratory, Stevenage)

Weathering of Building Stone and other Non-metallic Related Materials
Prof. R. U. Cooke (University College London) and Dr R. N. Butlin (Building Research Establishment, Garston; Royal Society of Chemistry)

Weathering of Metals and Coatings
G. O. Lloyd (National Physical Laboratory, Teddington)

Session 2 Chairman: Dr J. H. Chesters

Control Technologies and Remedial Strategies

C. J. Davies (National Coal Board, Harrow; Operational Research Society)

European Legislative Outlook
A. J. Clarke (Central Electricity Generating Board, London)

Emission Projections to 2000
Dr J. Skea (Science Policy Research Unit, University of Sussex)

Technologies for Stationary Sources
Dr D. Cope (IEA Coal Research, London; Institute of Energy)

Technologies for Mobile Sources
Dr J. Weaving (Institution of Mechanical Engineers)

Monitoring Implications
M. J. Woodfield (Warren Spring Laboratory, Stevenage)

Implications for Industry in the United Kingdom
M. J. Flux (Confederation of British Industry, London)

Liming Lakes
Dr G. D. Howells (Central Electricity Generating Board, Leatherhead; Institute of Biology)

Cost-Effective Abatement Policies
C. J. Davies

Session 3 Chairman: Prof. K. Mellanby, CBE

Discussion

Conclusions: Sir John Mason

THE WATT COMMITTEE ON ENERGY

Objectives, Historical Background and Current Programme

1. The objectives of The Watt Committee on Energy are:

   (a) to promote and assist research and development and other scientific or technical work concerning all aspects of energy;
   (b) to disseminate knowledge generally concerning energy;
   (c) to promote the formation of informed opinion on matters concerned with energy;
   (d) to encourage constructive analysis of questions concerning energy as an aid to strategic planning for the benefit of the public at large.

2. The concept of the Watt Committee as a channel for discussion of questions concerning energy in the professional institutions was suggested by Sir William Hawthorne in response to the energy price 'shocks' of 1973/74. The Watt Committee's first meeting was held in 1976, it became a company limited by guarantee in 1978 and a registered charity in 1980. The name 'Watt Committee' commemorates James Watt (1736–1819), the great pioneer of the steam engine and of the conversion of heat to power.

3. The members of the Watt Committee are 61 British professional institutions. It is run by an Executive on which all member institutions are represented on a rota basis. It is an independent voluntary body, and through its member institutions represents some 500 000 professionally qualified people in a wide range of disciplines.

4. The following are the main aims of the Watt Committee:

   (a) To make practical use of the skills and knowledge available in the member institutions for the improvement of the human condition by means of the rational use of energy;
   (b) to study the winning, conversion, transmission and utilisation of energy, concentrating on the United Kingdom but recognising overseas implications;
   (c) to contribute to the formulation of national energy policies;
   (d) to identify particular topics for study and to appoint qualified persons to conduct such studies;
   (e) to organise conferences and meetings for discussion of matters concerning energy as a means of encouraging discussion by the member institutions and the public at large;
   (f) to publish reports on matters concerning energy;
   (g) to state the considered views of the Watt Committee on aspects of energy from time to time for promulgation to the member institutions, central and local government, commerce, industry and the general public as contributions to public debate;
   (h) to collaborate with member institutions and other bodies for the foregoing purposes both to avoid overlapping and to maximise cooperation.

5. Reports have been published on a number of topics of public interest. Notable among these are *The Rational Use of Energy* (an expression which the Watt Committee has always preferred to 'energy conservation' or 'energy efficiency'), *Energy Development and Land in the United Kingdom*, *Energy Education Requirements and Availability*, *Nuclear Energy* and *Acid Rain*. Others are in preparation.

6. Those who serve on the Executive, working groups and sub-committees or who contribute in any way to the Watt Committee's activities do so in their independent personal capacities without remuneration to assist with these objectives.

7. The Watt Committee's activities are co-ordinated by a small permanent secretariat. Its income is generated by its activities and supplemented by grants by public, charitable, industrial and commercial sponsors.

8. The latest Annual Report and a copy of the Memorandum and Articles of Association of The Watt Committee on Energy may be obtained on application to the Secretary.

Enquiries to:
The Information Officer,
The Watt Committee on Energy,
Savoy Hill House,
Savoy Hill, London WC2R 0BU
Telephone: 01–379 6875

Member Institutions of The Watt Committee on Energy

British Association for the Advancement of Science

*British Nuclear Energy Society

British Wind Energy Association

Chartered Institute of Building

Chartered Institute of Building Services Engineers

*Chartered Institute of Management Accountants

*Chartered Institute of Transport

Combustion Institute (British Section)

Geological Society of London

Hotel Catering and Institutional Management Association

Institute of Biology

Institute of British Foundrymen

Institute of Ceramics

Institute of Chartered Foresters

*Institute of Energy

Institute of Home Economics

Institute of Hospital Engineering

Institute of Internal Auditors (United Kingdom Chapter)

Institute of Management Services

Institute of Marine Engineers

Institute of Mathematics and its Applications

Institute of Metals

*Institute of Petroleum

Institute of Physics

Institute of Purchasing and Supply

Institute of Refrigeration

Institute of Wastes Management

Institution of Agricultural Engineers

*Institution of Chemical Engineers

*Institution of Civil Engineers

Institution of Electrical and Electronics Incorporated Engineers

*Institution of Electrical Engineers

Institution of Electronic and Radio Engineers

Institution of Engineering Designers

*Institution of Gas Engineers

Institution of Geologists

*Institution of Mechanical Engineers

Institution of Mining and Metallurgy

Institution of Mining Engineers

Institution of Nuclear Engineers

*Institution of Plant Engineers

Institution of Production Engineers

Institution of Structural Engineers

International Solar Energy Society—UK Section

Operation Research Society

Plastics and Rubber Institute

Royal Aeronautical Society

Royal Geographical Society

*Royal Institute of British Architects

Royal Institution

Royal Institution of Chartered Surveyors

Royal Institution of Naval Architects

Royal Meteorological Society

Royal Society of Arts

*Royal Society of Chemistry

Royal Society of Health

Royal Town Planning Institute

*Society of Business Economists

Society of Chemical Industry

Society of Dyers and Colourists

Textile Institute

* Denotes permanent members of the Watt Committee Executive

Watt Committee Reports

The following Reports are available:

2. Deployment of National Resources in the Provision of Energy in the UK
3. The Rational Use of Energy
4. Energy Development and Land in the United Kingdom
5. Energy from the Biomass
6. Evaluation of Energy Use
7. Towards an Energy Policy for Transport
8. Energy Education Requirements and Availability
9. Assessment of Energy Resources
10. Factors Determining Energy Costs and an Introduction to the Influence of Electronics
11. The European Energy Scene
13. Nuclear Energy: a Professional Assessment
14. Acid Rain
15. Small-Scale Hydro-Power
17. Passive Solar Energy in Buildings
18. Air Pollution, Acid Rain and the Environment
19. The Chernobyl Accident and its Implications for the United Kingdom

For further information and to place orders, please write to:

ELSEVIER SCIENCE PUBLISHERS
Crown House, Linton Road, Barking, Essex IG11 8JU, UK

Customers in North America should write to:

ELSEVIER SCIENCE PUBLISHING CO., INC.
P.O. Box 1663, Grand Central Station, New York, NY 10163, USA

Index